High-Speed Precision Motion Control

High-Speed Precision Motion Control

Edited by
Takashi Yamaguchi
Mitsuo Hirata
Chee Khiang Pang

CRC Press
Taylor & Francis Group
Boca Raton London New York

CRC Press is an imprint of the
Taylor & Francis Group, an **informa** business

CRC Press
Taylor & Francis Group
6000 Broken Sound Parkway NW, Suite 300
Boca Raton, FL 33487-2742

First issued in paperback 2017

© 2012 by Taylor & Francis Group, LLC
CRC Press is an imprint of Taylor & Francis Group, an Informa business

No claim to original U.S. Government works

Version Date: 20111026

ISBN 13: 978-1-4398-6726-6 (hbk)
ISBN 13: 978-1-138-07160-5 (pbk)

Visit the Taylor & Francis Web site at
http://www.taylorandfrancis.com

and the CRC Press Web site at
http://www.crcpress.com

Contents

5 Ultra-Precise Position Control   137

Takenori Atsumi, Mituso Hirata, Hiroshi Fujimoto, and Nobutaka Bando

6 Control Design for Consumer Electronics   213

Mitsuo Hirata, Shinji Takakura, and Atsushi Okuyama

7 HDD Benchmark Problem   259

Mitsuo Hirata

List of Figures

List of Tables

Preface

This book describes high-speed precision motion control technologies which are developed and applied to hard disk drives (HDDs). The first feature of this book is that all the editors and the authors are engineers and professors who are directly engaged in the study and development of HDD servo control. Each author describes the control technologies that he developed, and most of these technologies have already been successfully applied to mass production of HDDs. As the proposed methodologies have been verified on commercial HDDs at the very least, these advanced control technologies can also be readily applied to precision motion control of other mechatronic systems, e.g., scanners, micro-positioners, photocopiers, atomic force microscopes (AFMs), etc.

The second feature of this book is that the control technologies are categorized into high-speed servo control, precision control, and environment-friendly control. As such, potential readers can easily find an appropriate control technology according to their domain of application. The control technologies described in this book also range from fundamental classical control theories to selected advanced topics such as multi-rate control.

Learning Outcomes
We expect this book to be useful to engineers, researchers, and students in technical junior colleges and universities as well as postgraduate students in various fields. Potential readers not working in the relevant fields can also appreciate the literature therein even without prior knowledge and exposure, and will still be able to apply the tool sets proposed to address realistic industrial problems. As such, engineers and managers are empowered with the knowledge and know-how to make important decisions and policies. Besides, this book can also be used to educate fellow researchers and the public about the advantages of various control technologies.

Many universities have established programmes and courses in this field, with much cross-faculty and inter-discipline research going on in this area as well. This book can also serve as a textbook for an intermediate to advanced module as part of control engineering, sampled-data systems, mechatronics, etc. We also hope that the book is concise enough to be used for self-study, or as a recommended text, for a single advanced undergraduate or postgraduate module on linear systems and digital control theory.

Acknowledgements

We would like to express our gratitude to university professors researching HDD servo control and HDD company engineers for their efforts in evolving high-speed precision motion control technology. We have learned a lot through various technical discussions and communications with all of them.

We would like to take this opportunity to express our gratitude to CRC Press for publishing this book. We would also like to acknowledge our loved ones for their love, understanding, and encouragement throughout the entire course of preparing this research monograph. This book was also made possible with the help of our colleagues, collaborators, as well as students, research staffs, and members of our research teams. This work was supported in part by Singapore MOE AcRF Tier 1 Grant R-263-000-564-133.

Last, but not least, we would like to take a moment to send all our best wishes to those who are affected, directly or indirectly, by the 2011 Eastern Japan great earthquake disaster.

Takashi Yamaguchi
Mitsuo Hirata
Chee Khiang Pang

MATLAB® is a registered trademark of The MathWorks, Inc. For product information, please contact:

The MathWorks, Inc.
3 Apple Hill Drive
Natick, MA 01760-2098 USA
Tel: +1 508 647 7000
Fax: +1 508 647 7001
E-mail: info@mathworks.com
Web: www.mathworks.com

About the Editors

Dr. Takashi Yamaguchi graduated from the Tokyo Institute of Technology with an M.S. in 1981. He joined the Mechanical Engineering Research Laboratory (MERL), Hitachi Ltd., in 1981, and worked on research and development of servo control of hard disk drives (HDDs) from 1987 to 2008. He received his Dr Eng. in 1998, and the title of his dissertation was *Study of Head Positioning Servo Control for Hard Disk Drives*.

Over the past thirty years, Dr. Yamaguchi's main research interests and areas have focused on motion control design, especially fast and precise positioning servo control design for HDDs. He has published 42 full papers, 26 articles and survey papers, 4 books as co-author, 71 presentations, and holds 28 US patents. Most of the publications are related to servo control of HDDs.

In 2008, he joined the Core Technology Research Center, Research & Development Group, Ricoh Company Ltd., where he is currently a general manager and an executive engineer. He is a fellow of the Japan Society of Mechanical Engineers (JSME) and a senior member of the Institute of Electrical Engineers in Japan (IEEJ).

He was a chief editor of *Nanoscale Servo Control*, TDU Press, 2007, which was the first book in Japan regarding the modelling and the control of HDDs. He was a guest editor for a special issue on "Servo Control for Data Storage and Precision Systems," *Mechatronics*, 2010.

Professor Mitsuo Hirata received his Ph.D. from Chiba University in 1996. From 1996 to 2004, he was a research associate of electronics and mechanical engineering at Chiba University. Currently, he is an associate professor of electrical and electronic systems engineering at Utsunomiya University.

Prof. Hirata has extensive research experience in the design and implementation of advanced control algorithms for mechatronic systems. Some past related projects include high speed and high precision control of head actuators of HDDs, semiconductor manufacturing systems (a collaboration with Canon Inc.), Galvano scanner (a collaboration with Canon Inc.), and transmission of vehicles (a collaboration with Nissan Motor Co., Ltd.), etc.

He was the co-author and editor of *Nanoscale Servo Control*, TDU Press, 2007, which was the first book in Japan regarding the modelling and the control of HDDs. The book includes an HDD benchmark problem in the attached CD-ROM, and he is the chair of a technical working group of the HDD benchmark problem that can also be obtained from the following URL: http://mizugaki.iis.u-tokyo.ac.jp/nss/MSS_bench_e.htm. He has published many international refereed journals and conference papers relevant to the scope of this book.

Professor Chee Khiang Pang, Justin, was born in Singapore in 1976. He received B.Eng. (Hons.), M.Eng., and Ph.D. degrees in 2001, 2003, and 2007, respectively, all in electrical and computer engineering, from the National University of Singapore (NUS), working closely with A*STAR Data Storage Institute (DSI), Singapore. In 2003, he was a visiting fellow in the School of Information Technology and Electrical Engineering (ITEE), the University of Queensland (UQ), St. Lucia, QLD, Australia, working on a probabilistic small signal stability of large-scale interconnected power systems project funded by the Electric Power Research Institute (EPRI), Palo Alto, California, USA. From 2006 to 2008, he was a researcher (tenure) with Central Research Laboratory, Hitachi Ltd., Kokubunji, Tokyo, Japan. In 2007, he was a visiting academic in the School of ITEE, UQ, St. Lucia, QLD, Australia, and was invited by IEEE Queensland Section to deliver a seminar. From 2008 to 2009, he was a visiting research professor in the Automation & Robotics Research Institute (ARRI), the University of Texas at Arlington (UTA), Fort Worth, Texas, USA. Currently, he is an assistant professor in the Department of Electrical and Computer Engineering (ECE), NUS, Singapore. He is a faculty associate with A*STAR DSI and a senior member of IEEE.

Prof. Pang is the author of *Intelligent Diagnosis and Prognosis of Industrial Networked Systems* (CRC Press, 2011). In recent years, he served as a guest editor for the *International Journal of Systems Science, Journal of Control Theory and Applications*, and *Transactions of the Institute of Measurement and Control*. He is currently serving as an associate editor for *Transactions of the Institute of Measurement and Control*, on the editorial board for *International Journal of Computational Intelligence Research and Applications*, and on the conference editorial board for IEEE Control Systems Society (CSS). He was listed in *Marquis Who's Who in the World*, 27th Edition, USA, 2010, and was the recipient of the Best Application Paper Award in the 8th Asian Control Conference (ASCC 2011), Kaohsiung, Taiwan, 2011.

List of Contributors

Hidehiko Numasato
Hitachi Global Storage Technologies Japan, Ltd.
Fujisawa, Kanagawa, Japan

Hiroshi Uchida
Hitachi Global Storage Technologies Japan, Ltd.
Fujisawa, Kanagawa, Japan

Shinji Takakura
Toshiba Corp.
Kawasaki, Kanagawa, Japan

Takeyori Hara
Toshiba Corp.
Ome, Tokyo, Japan

Dr. Nobutaka Bando
Japan Aerospace Exploration Agency
Sagamihara, Kanagawa, Japan

Dr. Takashi Yamaguchi
Ricoh Company Ltd.
Yokohama, Kanagawa, Japan

Dr. Takenori Atsumi
Hitachi, Ltd.
Fujisawa, Kanagawa, Japan

Prof. Atsushi Okuyama
Tokai University
Hiratsuka, Kanagawa, Japan

Prof. Hiroshi Fujimoto
The University of Tokyo
Kashiwa, Chiba, Japan

Prof. Mitsuo Hirata
Utsunomiya University
Utsunomiya, Tochigi, Japan

Nomenclature

A/D	Analog-to-Digital
AFC	Acceleration Feedforward Control
AFM	Atomic Force Microscopy
ARE	Algebraic Riccati Equation
BER	Bit Error Rate
CACSD	Computer-Aided Control System Design
CD	Compact Disc
CPU	Central Processing Unit
D/A	Digital-to-Analog
DFT	Discrete Fourier Transform
DOF	Degree-of-Freedom
DSP	Digital Signal Processor
DVD	Digital Versatile Disc
EMF	Electro-Motive Force
FB-RPTC	FeedBack Repetitive Perfect Tracking Control
FF-RPTC	FeedForward Repetitive Perfect Tracking Control
FFSC	Frequency-Shaped Final-State Control
FFT	Fast Fourier Transform
FIR	Finite Impulse Response
FIV	Flow-Induced Vibration
FSC	Final-State Control
HDD	Hard Disk Drive
HGST	Hitachi Global Storage Technologies
IDR	Inter-Sample Disturbance Rejection
IEEJ	Institute of Electrical Engineers of Japan
IIR	Infinite Impulse Response
ISS	Initial Shock Spectrum
IVC	Initial Value Compensation
LCD	Liquified Crystal Display
LMI	Linear Matrix Inequality
LPF	Low Pass Filter
LQG	Linear Quadratic Gaussian
LQR	Linear Quadratic Regulator
LTI	Linear Time-Invariant

MD	Mini Disc
MIMO	Multi-Input-Multi-Output
MPES	Master Position Error Signal
MPVT	Minimizing Primary Vibration Trajectory
MSC	Mode Switching Control
NRRO	Non-Repeatable Run-Out
ODOF	One-Degree-of-Freedom
OTC	Off-Track Capability
PCB	Printed Circuit Board
PCF	Phase Compensating Filter
PES	Position Error Signal
PID	Proportional–Integral–Derivative
PSG	Periodic Signal Generator
PTC	Perfect Tracking Control
PTOS	Proximate Time-Optimal Servomechanism
R/W	Read/Write
RLS	Recursive Least Squares
rpm	revolutions-per-minute
RPTC	Repetitive Perfect Tracking Control
RRO	Repeatable Run-Out
RSS	Residual Shock Spectrum
SAM	Servo Address Mark
SISO	Single-Input–Single-Output
SP	Sound Pressure
SPES	Slave Position Error Signal
SQP	Sequential Quadratic Programming
SRS	Shock Response Spectrum
STW	Servo Track-Writing
TDOF	Two Degrees-of-Freedom
TMR	Track Mis-Registration
TP	Track Pitch
TPI	Tracks-Per-Inch
VCM	Voice Coil Motor
ZOH	Zero Order Hold
ZPE	Zero-Phase Error
ZPETC	Zero-Phase Error Tracking Control

Chapter 1

Introduction

Takashi Yamaguchi

Ricoh Company Ltd.

1.1   Concept of High-Speed Precision Motion Control

First of all, it is important to define the title of this book "High-Speed Precision Motion Control." For accurate servo-positioning of mechanical actuators in realistic engineering systems, high quality motion is required to achieve both high speed and high precision positioning. As such, the typical four control systems design phases are:

1. design of reference trajectory;

2. design of controller to track the reference trajectory;

3. design of transient or settling controller to minimize the tracking error caused by various unmodeled dynamics or unpredicted plant fluctuations; and

4. design of controller to suppress external disturbances to ensure the controlled object remains on the target position.

To be more specific, the word "precision" must also be properly defined. A well-known metric for precision is the ratio between accuracy (or resolution) and stroke (or range). For high-performance positioning systems, ultra-high precision is usually in the order of magnitudes of 10^{-6} to 10^{-7} or less.

Many devices and equipment require high-speed precision motion control in industrial engineering systems. For example, the Hard Disk Drive (HDD) is one such unique device that requires high-speed precision motion control of the magnetic Read/Write (R/W) heads to perform read and write operations of user data on the magnetic disks.

The technologies required to achieve high-speed and precise positioning depend on whether the controlled variables such as position can be directly detected. In the case where the controlled variables can be directly detected, the control methodology is known as *full* closed-loop control. Otherwise, it is known as *semi* closed-loop or open-loop control. In the latter (which is a more popular method in industries due to the difficulty in selecting suitable sensors to detect the controlled variables), design efforts to achieve high precision are focused on keeping the operating conditions constant so that the controlled variables are not affected by unobservable external disturbances.

In the case of full closed-loop control, disturbances and plant dynamics as well as their fluctuations are included in the servo control loop, which causes the control system design to be much more challenging. However, once a satisfactory control loop can be designed, this method essentially has the potential to achieve the required precise positioning, since the errors between the reference and the controlled variables due to disturbances or fluctuations can be detected and minimized accordingly. From the viewpoint of control systems design, the full closed-loop control design methodology is more ideal and preferred. The head-positioning servo control of the R/W head in an HDD is one of full closed-loop control, since the control variable, i.e., the position error between the R/W head and the written data track, can be measured or detected directly. Traditionally, there have been many setbacks when designing controllers in order to realize the advantage of full closed-loop control in the history of HDD development. Subsequently, a positioning accuracy of several nanometers can be achieved under normal operating conditions in today's HDDs.

The detailed features of high-speed precision motion control described in this book are as follows:

1. **Control systems design based on the four control systems design phases**

    It is important to design the correct handover from high-speed motion control to precision motion control. Currently, many industrial controllers used in various engineering disciplines have two or more control modes, and a supervisory controller is commonly employed to switch between the tracking mode to the positioning mode in order to accomplish a given command such as moving and settling the controlled object to the target position. Each control mode is also usually designed to optimally meet the local performance index. For example, the performance index may be minimum time in the tracking mode, while disturbance suppression capability may be the performance index in the positioning mode. In Chapter 3, fundamental controller designs based on classical control theories and their extensions are described, including the entire structure of the proposed Mode Switching Control (MSC) framework. In Chapter 4, several ultra-fast motion control design methods based on advanced control theories are described, and several ultra-precise position control designs based on advanced control theories are discussed in Chapter 5.

2. **Control systems design considering control input saturation**
   When the distance from the current location to the desired target is sufficiently large, a maximum control input which saturates the power amplifier during acceleration is effective in shortening the actuation time. In this case, the design issue is how to track the controlled object on the reference trajectory precisely during deceleration, i.e., after releasing the control input of the power amplifier from saturation during acceleration (see Chapter 3.2.2).

3. **Control systems design considering vibration characteristics**
   One of the factors which deteriorates high-speed precision positioning is the vibration of the controlled object whose modes are easily excited by external disturbances or control input. This is especially true in the case of full closed-loop control, which takes into account *all* the vibration modes including those above the sampling frequency. It is also desirable to design the reference trajectory and its tracking control so as not to excite the vibration modes during actuation (see Chapters 3.4.1, 4.1, 5.1, and 5.2).

4. **Control systems design considering disturbance suppression capabilities**
   One of the most important indices in precise motion control design is the improvement in disturbance suppression capabilities. This generalizes to the demand for high servo bandwidth control and advanced sensor fusion techniques to detect disturbances so that corresponding disturbance rejection control methods can be used to suppress the detected external disturbances (see Chapters 5.1, 5.4, and 5.5).

5. **Sampling frequency selection for signal detection**
   It is desirable to detect controlled variables directly for precision motion control. The quality of the detected signals, such as noise level, resolution of the detected signal, and linearity, etc., are important performance measures. Moreover, as the sampling frequency of detected signals affects servo control performance, it is also necessary to develop control design methods to improve servo control performance using a specific sampling frequency (see Chapters 3.4.2, 4.2, 5.3).

6. **Influences on environment**
   With the current environmental concerns for sustainability and energy efficiencies, it is important to take into account environment factors such as power consumption and noise. A couple of approaches to design controllers with less influence and impact on the environment are included (see Chapters 6.1, 6.2, and 6.3).

The six items mentioned above are the main features of high-speed precision motion control covered in this book. The applications of these control systems design methodologies cover many industrial applications such as

XY-stage for semiconductor or Liquified Crystal Display (LCD) manufacturing equipment, ultra-precise measurement tools such as the three-dimensional probe, nanopositioning stages, Atomic Force Microscopes (AFMs), information storage devices such as HDDs, Compact Discs (CDs), and Digital Versatile Discs (DVDs), as well as positioning systems for medical and biological applications, etc.

In this book, the HDD is chosen as an industrial application example, and the servo control design methods as well as their effectiveness are verified. HDD control is selected due to the following reasons. First, HDD servo control is a simple Single-Input-Single-Output (SISO) system, where the magnetic R/W head is moved to a target data track in miliseconds and positioned on the track with nanometer accuracy. This is in essence high-speed precision control. Second, HDD servo control is a full closed-loop system that achieves the required positioning accuracy under various disturbance sources from the external environment of the HDD. Third, the HDD has many mechanical vibration modes in its feedback loop which must be kept stable using advanced vibration control techniques, as the magnetic R/W head is supported by a very complicated suspension mechanism which should ideally be rigid in the direction of actuation (in-plane) but flexible in the vertical direction (out-of-plane) to keep a small distance of less than 5 nm between the head and the disk surface. Last, HDD servo uses digital control which includes various design methodologies to achieve the required positioning accuracy with a relatively large sampling period for detecting the controlled variable or position of the R/W head.

In all, it would be beneficial for readers to understand high-speed precision motion control design systematically by learning actual HDD servo control design methods which have the above-mentioned features.

1.2   Hard Disk Drives (HDDs)

The picture of a commercial HDD is shown in Figure 1.1. In an HDD, one or more disks are stacked on the spindle motor shaft and rotate typically at 15,000 revolutions-per-minute (rpm) in high-performance HDDs and 5,400–7,200 rpm in mobile or desktop HDDs. Several hundred thousand data tracks are magnetically recorded on the surface of the disk with a track center-to-center spacing of less than 100 nm. The magnetic R/W head is mounted on a slider, which is in turn supported by the suspension and the carriage. The separation between the head and the disk is maintained by a hydrodynamic bearing. An electromagnetic actuator known as the Voice Coil Motor (VCM) rotates the entire carriage assembly and positions the slider on a desired track. The moving portion of the plant, i.e., the controlled object, consists of the VCM, carriage, suspensions, and sliders. The control algorithms are im-

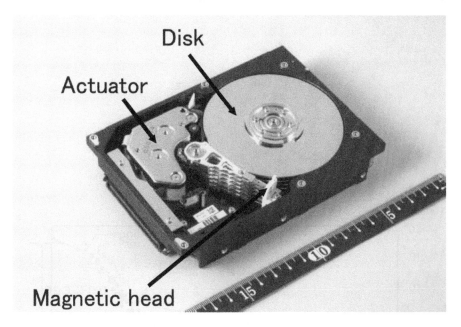

FIGURE 1.1: Schematic apparatus of a commercial HDD.

plemented in a Digital Signal Processor (DSP) or a microprocessor, which is mounted on a Printed Circuit Board (PCB).

The basic function of the HDD is to store and retrieve user data on a disk using only one magnetic R/W head, which makes its cost very low. However, this concept requires a good motion controller to move the head to the target track and position it on the track. Currently, the positioning accuracy is less than 10 nm at 3σ Non-Repeatable Run-Out (NRRO) in an HDD where σ is the standard variation.

The first shipment of HDDs was in 1956 from IBM. The capacity of HDDs then was 5 MB, and an HDD consisted of fifty disks each 24" in diameter. The disk rotation speed was 1,200 rpm, and the areal recording density was 2,000 bit/in^2 then. In 2006, a typical 2.5" HDD had 80 GB storage capacity per disk and its areal density is about 130 Gbit/in^2!

The trend of areal recording density in HDDs is shown in Figure 1.2 [1]. It can be seen that the the areal density has increased by 100 million times over the past 55 years! Since 2006, perpendicular recording has also been applied to actual HDDs, and it is expected to further increase the areal densities of HDDs in the future. At the same time, the positioning accuracy and speed of accessing the data tracks should also be improved. In 2010, a 2.5" HDD has a storage capacity of 250 GB per disk, and its areal density is more than 300 Gbit/in^2.

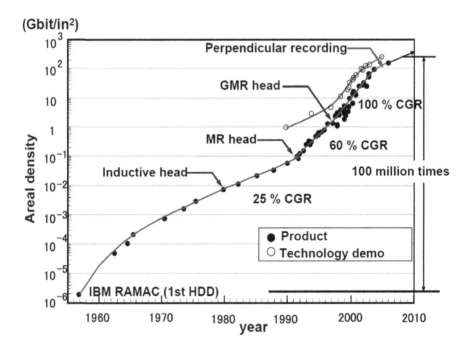

FIGURE 1.2: Trend of areal densities of HDDs.

TABLE 1.1: Short History of Servo Control Technologies Applied to HDDs

		Time (ms)	Trajectory design	Tracking control	Settling control
1970s				**Velocity control (1974)(Sec.3.2)**	Mode switching control(1977)
1980	5.25"				
	3.5"				
1985					
	First 3.5" HDD using digital servo				
				PTOS(Sec.3.2)	
			Jerk minimum trajectory(Sec.3.2)		
	2.5"	15		Finite settling control	**IVC(initial value compensation)*** **(Sec.3.3)**
1990			Input shaping	Sliding mode control	
	1.8"/1.3"	11		TDOF(Sec.3.2)	
		9		Model following control	Vibration control
1995		8			
					FF type IVC* **(Sec.3.3)**
		6		**N-delay multi-rate feedforward***(Sec.6.2)	
	1.0"			**Multi-rate PTC*** **(Sec.4.2)**	
2000				Deadbeat control	
		4		**Final state control*** **(Sec.4.1)**	
		3.5			**Residual vib. control using SRS analysis*** **(Sec.6.3)**
			Frequency shaped vib. suppression* **(Sec.4.1)**	Vib. suppression PTC IVC seek*(Sec.6.1)	Mode switching control considering control input restriction
2005	0.85"				

TABLE 1.2: Short History of Precision Control Technologies Applied to HDDs

			Precision control	
		Time (ms)	Robust control	Disturbance suppression, adaptive control
1970s			**Phase compensation(1974)(Sec.3.4)** **Kalman filter(1976)(Sec.3.4)**	
1980	5.25"			
	3.5"			
1985				
	First 3.5" HDD using digital servo			
			Observer based state feedback (Sec.3.4)	Disturbance observer
			LQG/LTR	Adaptive RRO compensation
	2.5"	15		**Accel. Feedforward control (Sec.5.5)**
1990			**Multi-rate control(Sec.3.4)**	**Repetitive control(Sec.5.4)**
	1.8"/1.3"	11	Two stage actuator	Self-tuning control
			H$_\infty$ control*(Sec.5.2)	
		9		Learning control
			H$_2$ control	Friction control
1995		8	Multi-rate notch filter	Adaptive seek control
			Multi sensing control	
		6	Mechanical resonance in-phase control PQ control(decoupling)	RRO compensation (peak filter, feedforward, and non-tracking)
	1.0"		Actuator-sensor collocation control	
2000			**Sampled H$_\infty$ control*(Sec.5.3)** Active damping control	MRACS
		4		
				Adaptive model identification accel. feedforward* (Sec.5.5)
		3.5		**Multi-rate RRO control* (Sec.5.4)**
			Mechanical resonance phase stable control *(Sec.5.1) Control using virtual resonant mode	RRO compensation (Inverse transfer function)
2005	0.85"			Hybrid identification

Tables 1.1 and 1.2 show the history of control theories and technologies that were applied to or to be applied to HDDs since the 1970s, when digital control was first introduced to HDDs. This information was collected from various sources reported in international technical journals and conference proceedings. The technologies in bold are described in future chapters, and the technologies with '*' are proposed by one of the authors of this book. It can be seen from Tables 1.1 and 1.2 that for many years, engineers have made many research efforts to apply the latest advanced control theory developments to HDD servo control.

Bibliography

[1] R. Wood and H. Takano, "Prospects for Magnetic Recording over the Next 10 Years," in *Digest of the IEEE INTERMAG*, pp. 98, 2006.

Chapter 2

System Modeling and Identification

Hiroshi Uchida

Hitachi Global Storage Technologies Japan, Ltd.

Takashi Yamaguchi

Ricoh Company Ltd.

Hidehiko Numasato

Hitachi Global Storage Technologies Japan, Ltd.

2.1 HDD Servo Systems

In this section, the Hard Disk Drive (HDD) servo system is introduced. The components of a typical HDD that constitute the HDD servo system are first presented, followed by concepts of servo pattern and servo position signal generation techniques [1, 2, 3].

2.1.1 Inside an HDD

A typical Hard Disk Drive (HDD) consists of hard disk platters where information or user data is magnetically recorded, as shown in Figure 2.1. The HDD contains:

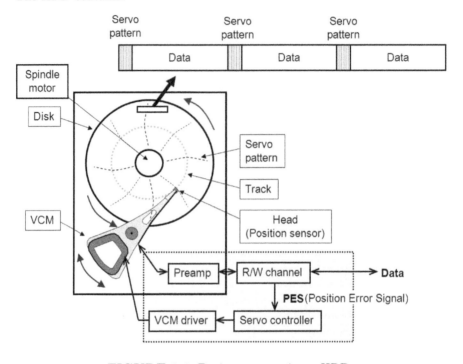

FIGURE 2.1: Basic structure in an HDD.

1. disks which are rotated by a spindle motor;

2. magnetic Read/Write (R/W) heads which are magneto-resistive sensors for reading user data and servo position signals from the disks and writing user data onto the disks;

3. a Voice Coil Motor (VCM) which actuates and positions the head assembly;

4. a pre-amplifier and signal processing circuit which reads back and demodulates the magnetic signal or modulates and writes the magnetic signal; and

5. a servo controller and VCM driver circuit which ensures that the R/W head seeks and follows the data track, etc.

Data is recorded in concentric circles on the disks using the magnetic head, and one such circle is known as a *data track* or simply track. Consecutive numbers in an ascending sequence known as the *cylinder number* are allocated to all tracks. In a track, consecutive numbers known as the *data sector number* (or sector number) are allocated in the direction of the disk rotation. Similarly, the *servo sector number* is also allocated to every servo pattern block in an ascending order. In addition, consecutive numbers known as the *head number* are allocated to each R/W head. User data is then recorded to or retrieved from the location uniquely addressed by its cylinder number, head number, and data sector number.

In the data track on a disk, magnetic patterns known as *servo patterns* are recorded on the disks (similar to grooves in a record disc) during the *Servo Track-Writing* (STW) process *a priori*. Servo position signals are then available at a fixed sampling time, and Position Error Signals (PES) are then generated from the servo signal for position control of the head to achieve precise track-following.

The track width in the latest versions of HDDs is less than 150 nm, i.e., 0.15 μm. To ensure that user data on the adjacent track is not erased during writing, the 3σ of head-positioning accuracy should be less than 10% of the track width. In some mobile HDD applications, the positioning error is even less than 10 nm. As such, nanopositioning control is required and performed in HDDs.

2.1.2 Generation of Servo Position Signal

In this section, the concepts of servo pattern and servo position signal generation techniques are explained.

An example of servo pattern and its read back signal is shown Figure 2.2. In the servo pattern, the grey areas show the locations where magnetic patterns are recorded, while the white areas show the locations where magnetic patterns are not recorded or erased. As the read element of the magnetic head passes through the position at a small offset from the track center, the corresponding read back signal is shown in the top figure. In the servo pattern, there are several areas known as the preamble, Servo Address Mark (SAM), servo sector number, servo track number, and A, B, C, and D servo bursts. Since the servo pattern is arranged differently from the data pattern in a magnetic sense, a servo signal detection circuit is used to synchronize the read back signal at the servo pattern at the preamble. This is differentiated from a servo mark as the start of the servo pattern from the read back signal. The read back signal is then decoded into a servo track number and track or cylinder number. These numbers are digitally recorded using the Gray code with the hamming distance between adjacent tracks being always one. As such, the Gray code differs a bit between adjacent tracks, which makes it possible to detect a one bit error in the read back signal even at half-track position.

The read back signals at bursts A, B, C, and D in Figure 2.2 are used for calculating the PES within a track. To be more specific, the read back signals at bursts A, B, C, and D are multiplied by sine waves whose frequencies are

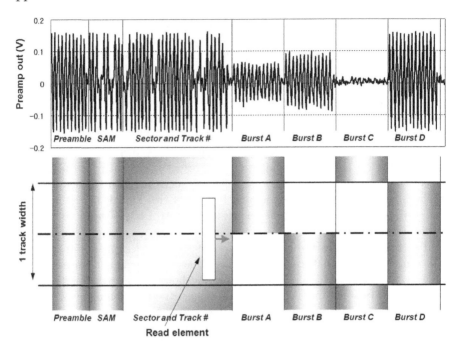

FIGURE 2.2: Read back signal (top) and servo pattern (bottom).

identical to the servo pattern and summed up. The added values are then used for calculating the *Master Position Error Signal* (*MPES*) and *Slave Position Error Signal* (*SPES*) using

$$MPES = \frac{A - B}{A + B}, \tag{2.1}$$

$$SPES = \frac{C - D}{C + D}. \tag{2.2}$$

The relationship between the servo pattern and PES is shown in Figure 2.3. Since *MPES* and *SPES* differ by 90° in phase, the signal with better linearity on position error is selected as the PES. This method is known as the *two-phase servo pattern*. In HDDs, the signals are compensated to improve linearity, which might not be always achievable due to the asymmetry of the read element's sensitivity and magnetization differences at the edges of servo patterns that occurred during the erase or overwrite process. In this way, the PES in HDDs are in orders of nanometers.

PES is an error signal between the target position and current head position measured by the coordinates of the servo pattern. As can be seen from the absolute coordinate, the target position is not an ideal concentric circle as the servo pattern is distorted due to disk flutter/disk vibration, disk eccentricities, and even the servo pattern itself, etc., which contains position error

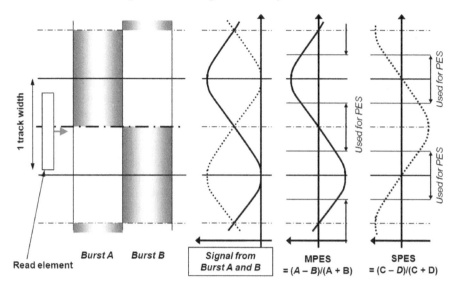

FIGURE 2.3: PES generation using burst signals read from the servo burst pattern.

due to vibration during the STW process. From the point of view of magnetic recording, it is very important to improve the accuracy of servo patterns during the STW process. On the other hand, the PES sensed by the read element of the head is just a controlled value which should be maintained at zero during servo control. In HDDs, the servo controller should minimize the *relative position error* between the position target and the current head position. Fortunately, a *full closed-loop control* can be realized since this signal can be detected directly as PES.

In order to describe an HDD model more accurately, the following concepts of:

1. current head position;

2. target position;

3. coordinate distortion of servo pattern due to disturbances; and

4. position error between the current head position and the target position measured by a distorted coordinate

are explained in this chapter. In future chapters, the position error due to a distorted coordinate is treated as a *position disturbance* and is added to the output before sensing the PES. Since PES is a controlled variable and is detectable, it can be defined by the difference between the position signal y and the target position set forth by the reference r which is similar to conventional servo control approaches.

2.2 TMR Budget Design

It is important to design a head-positioning servo system under the supervision of the systematic TMR budget design for HDDs. In general, there are several levels of the TMR budget design [4]. The first level involves a model that describes the relationship between the Bit Error Rate (BER) and the head-positioning error. The next level is a model describing TMR, which is defined as the *positioning error from the ideal position*, as shown in Figure 2.4. Since a magnetic head in HDD is always affected by disturbances such as air flow or windage which causes FIV, modeling of disturbance is as important as modeling of the plant dynamics. The last level is a detailed model describing the relationship between the positioning accuracy and servo-mechanical characteristics. In this section, a basic model for the TMR budget design is presented.

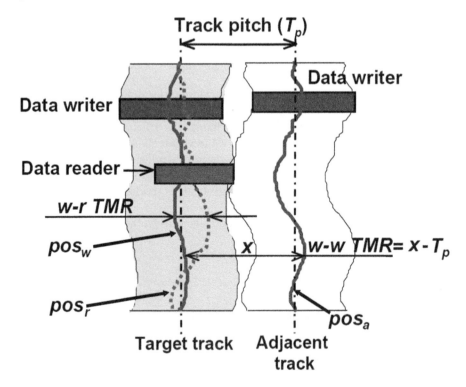

FIGURE 2.4: $w - wTMR$ and $w - rTMR$.

A basic flow of the TMR budget design is shown in Figure 2.5. When product specifications such as data capacity, data transfer rate, and number of disks, etc., are given, basic design specifications such as BER, areal density

Product specification
Data capacity, data transfer rate, etc.

Specification for product design
Areal density (track density, bit density), bit error rate, seek time, rotational speed, etc.

Specification of recording system
TMR (track misregistration) budget design at data read/write
w-wTMR, w-rTMR, head size

Position error factor model

Design specification for mechanical and servo control components
(1)Suppression of disturbance (windage etc.)
(2)Tracking capability (servo bandwidth etc.)
(3)Quality of position signal (accuracy of servo writing, S/N etc.)

FIGURE 2.5: Basic design flow of HDD head-positioning system.

consisting of track density, bit density, track-seeking time (minimum, average, and maximum moving time from track to track), and disk rotational speed, etc., can be obtained. Next, the size of the head and head-positioning error during data writing and reading are given based on the track density, after which the specification of positioning accuracy at each head during data writing and reading is given in terms of the head-positioning error. Finally, the mechanical vibration characteristics such as peak of resonant modes, gain of disturbances, servo control bandwidth, etc., are determined based on the required head-positioning accuracy.

The basic configuration representing the relationships among the self-track signal, adjacent track signal, and old information is shown in Figure 2.4. The BER (BER) in a specific position x is calculated using a modified error function curve based on experiments. Off-track capability (OTC) is defined as the width of a bathtub curve at a certain bit error rate such as 10^{-7}. The major HDD specification *average BER* is then calculated from OTC, $w - rTMR$, and $w - wTMR$ as

$$BER = f(w - rTMR, w - wTMR, OTC). \qquad (2.3)$$

The two TMRs, $w - rTMR$ and $w - wTMR$, are shown in Figure 2.4. $w - rTMR$ is the offset between the head trajectory $pos_w(t)$ writing the data and $pos_r(t)$ reading the data. $w - wTMR$ is an error between the nominal track pitch T_p and the distance between two adjacent trajectories of written data. Each trajectory is influenced by both head and disk vibrations as

$$
\begin{aligned}
w - rTMR(t) &= pos_w(t) - pos_r(t), & (2.4) \\
w - wTMR(t) &= |pos_w(t) - pos_a(t)| - T_p, & (2.5)
\end{aligned}
$$

where $pos_a(t)$ is the head trajectory from writing the data on an adjacent track. Both TMRs cannot be measured directly, but each head trajectory can be measured as a position error pe from the track center.

These TMRs are furthered classified into the following positioning errors. Since the position signal or servo pattern in HDDs is written by the STW process before shipping, this position signal also includes positioning error due to disturbances during STW. Each TMR consists of head position during data writing, head position during adjacent data writing, and head position during data reading. As such, it is necessary to define these position errors in detail. These position errors consist of fluctuation of servo signal which serves as a reference for head-positioning servo control, vibration of head due to various disturbances, and tracking error from feedback control. In all, these TMRs are:

1. error of position signal during STW, e.g., vibration during STW and static track pitch error;

2. error of position signal due to signal quality such as Signal-to-Noise Ratio (SNR) of position signal and nonlinearity;

3. fluctuation of position signal during data writing due to disk vibration;

4. vibration of head during data writing;

5. tracking error of head to servo signal during data writing;

6. fluctuation of position signal during data writing of adjacent track;

7. vibration of head during data writing of adjacent track;

8. tracking error of head to servo signal during data writing of adjacent track;

9. fluctuation of position signal during data reading due to disk vibration;

10. vibration of head during data reading; and

11. tracking error of head to servo signal during data reading.

The above-mentioned error factors are shown in Figures 2.6–2.10.

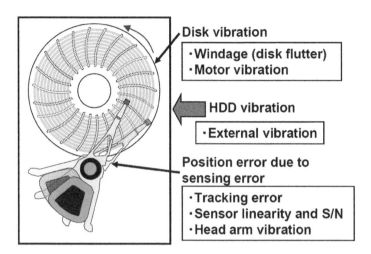

FIGURE 2.6: Error factor during position signal writing.

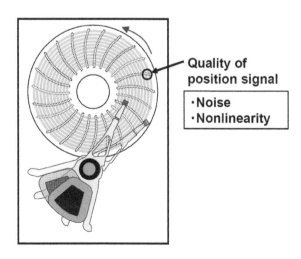

FIGURE 2.7: Error factor of position signal.

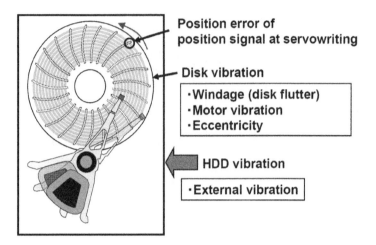

FIGURE 2.8: Error factor of position signal fluctuation during data reading/writing.

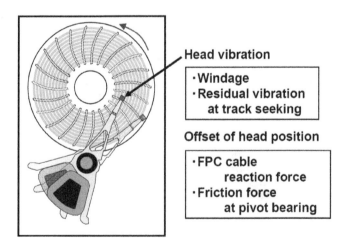

FIGURE 2.9: Error factor of head vibration during data reading/writing.

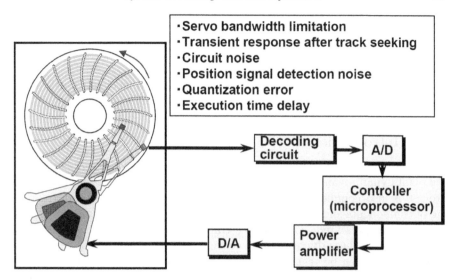

FIGURE 2.10: Error factor of tracking error during data reading/writing.

As these positioning errors are caused by the same phenomenon, the error factors should be rewritten concisely as:

1. quality of position signal;

2. vibration of position signal or disk during data writing and reading;

3. vibration of head during data writing and reading; and

4. tracking error of head to position signal during data writing and reading.

It should be noted that the "vibration of position signal or disk during data writing and reading" should be further classified into RRO such as disk eccentricity and NRRO. The RRO does not influence TMR since relative position among "fluctuation of position signal at data writing due to disk vibration," "fluctuation of position signal at data writing of adjacent track," and "fluctuation of position signal at data reading due to disk vibration" do not change. Strictly speaking, a portion of "error of position signal during STW, e.g., vibration at servo writing and static track pitch error" consists of RRO but this also does not influence TMR. Thus, the NRRO is essentially in "vibration of position signal or disk during data writing and reading" and should be taken into account in the design of the TMR budget. Consequently, the HDD head-positioning system has the following three design indices:

1. suppression of disturbances caused by mechanical vibrations;

2. tracking capability of servo control; and

3. accurate sensing devices.

These design indices would also be applicable to other motion control devices.

2.3 Modeling of HDD

In this section, the modeling of the components and mechanical dynamics of the head-positioning control system in HDDs are presented.

2.3.1 Introduction

The first and most important step for servo controller design is to identify the plant dynamics and describe them as a mathematical representation. This process is also commonly known as *plant modeling*. In this section, modeling of components and mechanical dynamics for the head-positioning systems in HDDs are presented. The Laplace transform with the *s*-operator will be used to express these dynamics in a continuous-time domain, and this representation can then be transformed to a state-space or discrete-time domain for time domain simulations as well as controller designs using modern and post-modern control theories.

2.3.2 Plant Components

The block diagram of the head-positioning system in HDDs is shown in Figure 2.11. In Figure 2.11, the time delay of the Digital-to-Analog (D/A)

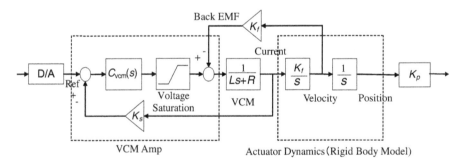

FIGURE 2.11: Plant block diagram.

converter and controller calculation are not explicitly shown but should be included in the plant model. The first-order *Pade approximation* is

$$G(s) = \frac{1 - \frac{\tau}{2}s}{1 + \frac{\tau}{2}s}, \qquad (2.6)$$

where τ is the total time delay as shown and $G(s)$ is one of the many ways used to represent this time delay [5]. The Pade approximation has an unstable zero and is a constraint for controller design in many cases.

The VCM driver for HDDs shown in Figure 2.11 usually has a current feedback loop which supplies voltage to the VCM coil based on the error between the input reference and the measured VCM current. The feedback loop is expected to compensate for the back Electro-Motive Force (EMF) caused by the movement of the VCM, and the compensator in the loop has a zero (phase-lead) to cancel the VCM pole at $-\frac{R}{L}$ (phase-lag) for extending the servo loop bandwidth. By increasing the bandwidth, the closed-loop transfer function of the VCM driver approximates the constant gain $\frac{1}{K_s}$ in an ideal case. However, in reality, the driver voltage is limited and the dynamics of the VCM are too complicated to be cancelled out by a zero of the compensator. These restrictions should be considered in the design of the current-feedback loop. The maximum velocity of the head-positioning system can be determined from the maximum driver voltage and VCM torque constant. In general, the VCM current feedback loop has to be designed to have:

1. higher bandwidth than the lowest natural frequency of the major mechanical mode; and

2. no large peak in gain in the closed-loop transfer function.

Under these conditions, the VCM driver can be approximately represented as a first-order lag given by

$$G_{mp}(s) = K_A \frac{\omega_{VCM}}{s + \omega_{VCM}}, \qquad (2.7)$$

where K_A is the DC gain and ω_{VCM} is the bandwidth or cut-off frequency of the VCM driver.

The head-positioning system of HDDs is an example of a full-closed loop system. The magnetic head reads the position signal, which is recorded at the servo sectors on the disk, and the position of the head motion is controlled based on this signal. The sampling period T_s in s is determined by the number of servo sectors in a track N_s and disk rotational speed N_r in revolutions per minute (rpm) given by the following relationship

$$T_s = \frac{60}{N_s \times N_r}. \qquad (2.8)$$

The integer and fraction parts of the position information data consist of the track number and shift from the track center, respectively, and can be measured from the change in amplitude of the burst signal. The non-linearity of the fraction parts of position information is well known [6], and it is usually compensated in HDDs. As such, the position detection function can be modeled as a gain K_{POS} under this assumption.

Based on these notions, the plant model can now be expressed as

$$P(s) = K_{POS} K_A \frac{1 - \frac{\tau}{2}s}{1 + \frac{\tau}{2}s} \frac{\omega_{VCM}}{s + \omega_{VCM}} M(s), \qquad (2.9)$$

where $M(s)$ represents the mechanical dynamics to be presented in the next section.

2.3.3 Modeling of Mechanical Dynamics

The conceptual diagram of the head-positioning mechanisms in HDDs is shown in Figure 2.12.

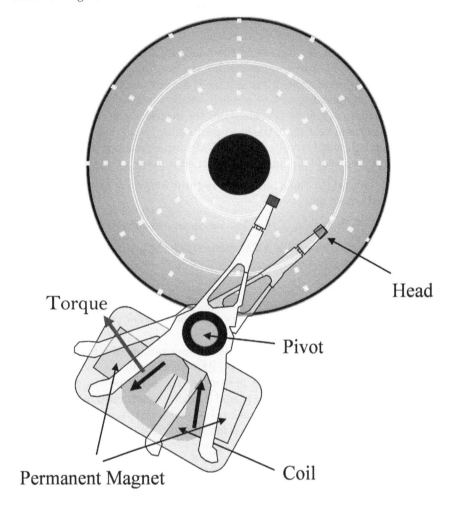

FIGURE 2.12: Head-positioning mechanisms in HDDs.

In HDDs, each head is attached to the tip of the suspension and carriage, and the VCM actuator moves the carriage, suspensions, and heads together in the radial direction. The VCM actuator consists of a permanent magnet, coil, and pivot, and the torque generated is proportional to the coil current. It is possible to transform the *rotational* motion equation to a *translational* motion equation. However, the gain of the plant should be compensated based on the yaw angle, which is given by the cosine of the angle between the radial direction and the tangent of the rotary motion of the head since the position signal is the displacement of the head in the radial direction.

In general, the modeling of mechanical resonant modes is performed based on matching both the gain and phase of the frequency response data measured from the VCM current to the head position. An example is shown in Figure 2.13. If there are no resonant modes, the model is known as a *rigid*

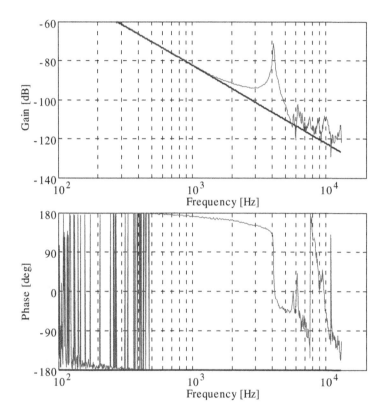

FIGURE 2.13: Measured actuator dynamics and rigid body model.

body with a -40 dB/dec gain slope and $-180°$ constant phase. As such, the difference from the rigid body characteristics arises due to the presence of mechanical resonant modes.

In general, there are two major formulae used to describe the mechanical dynamics, namely, the Σ-*type* expression connecting the resonant modes in *parallel* given by

$$M(s) = \frac{K_1}{s^2} + \sum_{r=1}^{N} \frac{1}{K_r} \frac{\omega_r^2}{s^2 + 2\zeta_r\omega_r s + \omega_r^2}, \qquad (2.10)$$

and the Π-*type* expression connecting the resonant modes in *series* given by

$$M(s) = \frac{K_1}{s^2} \times K_2 \prod_{r=1}^{N} \frac{s^2 + 2\zeta_{nr}\omega_{nr}s + \omega_{nr}^2}{s^2 + 2\zeta_{dr}\omega_{dr}s + \omega_{dr}^2} \tag{2.11}$$

$$= \frac{K_1 K_2(s^{2N} + b_{2N-1}s^{2N-1} + \cdots + b_1 s + b_0)}{s^2(s^{2N} + a_{2N-1}s^{2N-1} + \cdots + a_1 s + a_0)}. \tag{2.12}$$

In the Σ-type representation depicted in (2.10), K_1 is the gain of the rigid body mode, w_r is the angular frequency in rad/s, z_r is the mode damping ratio, and K_r is the equivalent stiffness. $\frac{1}{K_r}$ is known as the *residue* or *mode constant* [7][8]. In the Π-type representation depicted in (2.11), w_{nr} and z_{nr} are the natural frequencies and damping ratios of the anti-resonant modes, respectively. w_{dr} and z_{dr} are the natural frequencies and damping ratios of the resonant modes, respectively. a and b in the Π-type representation are coefficients of the transfer function. It is well known that the rigid body characteristic changes to a second-order delay element given by

$$\frac{K_1 \omega_0^2}{s^2 + 2\zeta_0 \omega_0 s + \omega_0^2} \tag{2.13}$$

in a small displacement range due to nonlinear friction. This is dependent on the amount of displacement and surfaces as *hysteresis* in time-domain responses [9].

For most cases in HDDs, w_0 is much lower than the mechanical resonant frequencies and servo bandwidth and hence the rigid body mode is a good approximation which suffices for servo control design. The Σ-type expression depicted in (2.10) is obtained based on a mechanical modal analysis technique assuming orthogonality properties of the eigenvectors of the vibration mode, and has the same modal parameters, i.e., angular frequency, damping ratio, and residues [7][8]. As such, the Σ-type model makes it easier for discussion on mechanical dynamics in a cross-functional setting involving mechanical and servo designers using these modal parameters. The modeling results using the Σ-type expression by adjusting three parameters for each model manually are shown in Figure 2.14 and Table 2.1 [10]. The bold line shows the simulation results using the Σ-type model and the thin line shows the measurement data. The model has six modes (a twelve-order transfer function) up to a frequency of 10 kHz. In general, the guidelines for constructing a Σ-type model are

1. begin with the modes with larger gain before proceeding to the modes with smaller gain;

2. set the modal angular frequency to match the frequency of the peak gain;

3. tune the modal damping ratio by using the half value method [7] in order to fit frequency-width of the peak gain at half of the maximum and shape of phase change; and

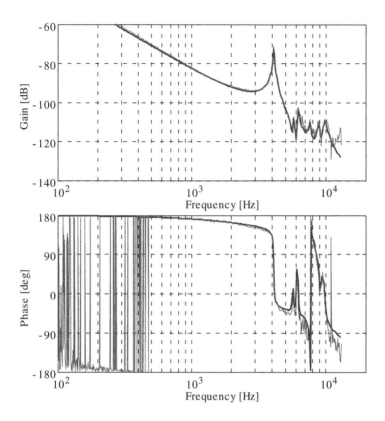

FIGURE 2.14: Modeling of actuator dynamics using the Σ-type model.

TABLE 2.1: Example of Mechanical Dynamics Modeling Using the Σ-Type Model

r	1	2	3
ω_r (rad/s)	$4100 \times 2\pi$	$5700 \times 2\pi$	$6200 \times 2\pi$
ζ_r	0.01	0.01	0.01
ω_r^2/K_r	-1.30	-0.03	-0.08
r	4	5	6
ω_r (rad/s)	$7650 \times 2\pi$	$8900 \times 2\pi$	$9800 \times 2\pi$
ζ_r	0.02	0.02	0.03
ω_r^2/K_r	0.12	-0.13	-0.35

4. adjust the absolute value and sign of the residue to match the peak gain
   and direction of phase change in the data, respectively.

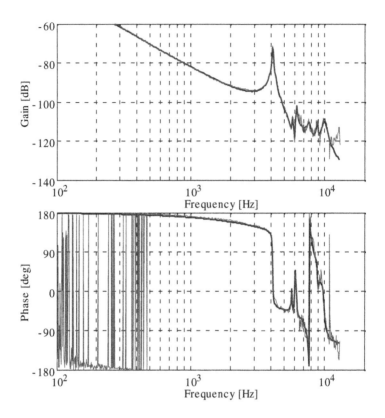

FIGURE 2.15: Modeling of actuator dynamics using the Π-type model.

The modeling results using the Π-type and the MATALB® command
`fitsys.m` are shown in Figure 2.15 and Table 2.2. The frequency response
data was transferred from *compliance* in Figure 2.13 to *inertance* by subtracting the rigid body characteristics and was used for identification. The order
of the plant was set at twelve to be the same as that used in the Σ-type modeling, and a weighting function shown in Figure 2.16 was used. The weighting
function had a baseline rigid body characteristic with a reduction in gain from
3.5 kHz to 5 kHz to avoid emphasis on the first resonant mode which has a
large gain. Larger weight gains are used on the small modes to achieve better
fitting.

The natural frequencies and damping ratios of both poles and zeros derived from the identified transfer functions are shown in Table 2.2. The parameters of the denominators in Tables 2.1 and 2.2 are very close, which implies
that both the Σ-type and Π-type modeling methods are able to represent

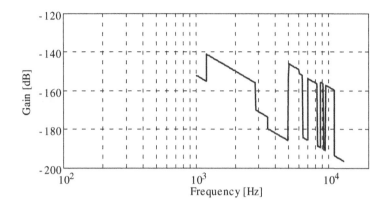

FIGURE 2.16: Weighting function used for Π-type modeling.

TABLE 2.2: Example of Mechanical Dynamics Modeling Using the Π-Type Model

r	1	2	3
ω_n (rad/s)	$5683 \times 2\pi$	$6040 \times 2\pi$	$7922 \times 2\pi$
ζ_n	0.010	0.009	-0.082
ω_d (rad/s)	$4109 \times 2\pi$	$5730 \times 2\pi$	$6167 \times 2\pi$
ζ_d	0.008	0.0081	0.008
r	4	5	6
ω_n (rad/s)	$8311 \times 2\pi$	$8924 \times 2\pi$	$9345 \times 2\pi$
ζ_n	0.237	0.024	-0.976
ω_d (rad/s)	$7719 \times 2\pi$	$8846 \times 2\pi$	$9997 \times 2\pi$
ζ_d	0.032	0.009	0.029

TABLE 2.3: Transformation of Zeros from Σ-Type Model to Π-Type Model

r	1	2	3
ω_n (rad/s)	$5634 \times 2\pi$	$6046 \times 2\pi$	$7802 \times 2\pi$
ζ_n	0.010	0.010	0.106
r	4	5	6
ω_n (rad/s)	$7834 \times 2\pi$	$9187 \times 2\pi$	$-5080 \times 2\pi$ $-5843 \times 2\pi$
ζ_n	-0.065	0.024	real zeros

the dynamics of the same plant well. A negative damping ratio in Table 2.2 indicates the presence of an unstable zero, as the `fitsys.m` command was executed without imposing the restriction of a minimum-phase system. The natural frequencies of the numerators ω_n and the anti-resonant frequencies of the frequency response also match well, as shown in Figure 2.13. When using the Π-type model, anti-resonant modes surface as zeros which block the transmission of input to output due to a low gain but the Σ-type model presents zeros as the points where modes cancel one another due to opposite phases.

It is possible to transform the Π-type model to the Σ-type model and vice versa. Table 2.3 shows the parameters of numerators using the Π-type model transformed from the Σ-type model in Table 2.1. By comparing the results in Tables 2.2 and 2.3, it can be verified that the natural frequencies and damping ratios match very well. However, the modes $r = 3$ and $r = 4$ should be switched considering the values of the damping ratios. In addition, the heavily damped mode $r = 6$ in Table 2.2 should replace the heavily damped mode in Table 2.3, though there are no significant differences in their frequency responses.

The results show that the Σ-type and Π-type models are essentially the same from a transfer function point of view (input-output characteristics), even when the parameters differ in terms of residues or zeros in the numerators. This notion is identical to the non-uniqueness of state-space equations when describing identical dynamical systems using different state variables. Either way, plant modeling is the first step before servo design, and the control systems designer has freedom of choice in terms of model representations, e.g., balanced-order representation from a numerical accuracy point of view, transfer function representation, state-space representation, etc., since these representations can be readily transformed from one to the other in later design phases.

There exists some trade-off between modeling accuracies, model orders, and work loads. In general, making a precise plant model requires equations of high-order and the work load required to accomplish the identification task is usually heavy. Also, high order models will cause numerical problems during controller design synthesis at times. As such, the following guidelines for plant modeling from a servo control design point of view are:

1. Mechanical modes which are not gain-stabilized by notch filters nor phase-stabilized by phase condition must be described precisely, since these modes will appear in the open loop transfer function with a large gain. The stability margins will also have to be rechecked using the phase condition [11][12];

2. Mechanical modes which are gain-stabilized are robust to modeling errors within the design margin of the notch filters. These modes do not appear in the open loop transfer function with a large gain peak and have almost no impact on the closed-loop transfer functions. It is also possible to model the combined frequency responses of the gain-stabilized modes and notch filters instead of modeling them separately;

3. Mechanical modes which are excited during track-seeking should be modeled precisely. Mechanical modes which are not excited can withstand some modeling errors.

From the point of view of notch filter design, it is unnecessary to represent the mechanical resonant modes using equations, since classical Bode and Nyquist plots are graphical methods which do not require mathematical representations. Stability margins can be checked from the gain and phase of open loop transfer functions in the frequency domain, and servo performance can be estimated from the shape of the closed-loop transfer functions. As such, the controller and notch filters can be designed in an iterative manner consisting of the following steps:

1. calculate the frequency response data of controller and notch filters;

2. combine the data with plant measurement data to calculate the open loop transfer function;

3. transform the open loop to a closed-loop transfer-function;

4. check the stability margins, servo bandwidth, and loop shape; and

5. if the results of Step 4 do not satisfy the design requirements, go back to Step 1 and follow the same procedures again using a new candidate for controller and/or notch filters

As an HDD consists of multiple heads, the mechanical resonant modes deviate from head-to-head and drive-to-drive and also vary with temperature. As such, much data is required for a good notch filter design to have sufficient robustness against these plant uncertainties. In this case, making mathematical representations for all the mechanical dynamics data is not feasible and the above-mentioned approach is more realistic for notch filter design. The modeling of plant model perturbation and modeling errors can be carried out using post-modern H_∞ and μ-synthesis control theories but it is also not an easy task [13]. If the thermal variation of a certain mechanical mode is steady and the mode has a large gain peak, redesigning the notch filter for that mode based on this information should be considered to reduce conservativeness.

It is worth noting that plant modeling and corresponding servo loop design are not one-time processes. During the design process, a more accurate modeling on some modes may be required, e.g., stability margins are critical at some frequencies, residual vibrations are large in time response simulations, etc., and iterations between modeling and servo design are common in the HDD industries.

The plant model of the head-positioning control system in HDDs consists of a second-order delay describing non-linear friction of the pivot or simply a rigid body double-integrator, coupled with computational delay and mechanical resonant modes. With the increase in positioning accuracy, more precise modeling is required which brings about the integration of servo and mechanical for synergetic modal analysis and modeling in the mechanical technical area and servo design. It is of paramount importance for the servo design process to identify the mechanical modes which limit servo bandwidth, and redesign each mode in order to phase-stabilize each mode via mechanical systems design. The HDD Benchmark Problem is presented in Chapter 7.

2.4 Modeling of Disturbances and PES

Positioning error in the steady-state is caused by disturbances which enter the feedback loop of the servo control systems. As such, *disturbance modeling* is required for evaluating its quantitative impact on the positioning error to enhance the performance. In this section, a disturbance modeling method which decomposes the PES into its contributing components, namely, the mechanical vibrations, PES demodulation noise, and torque noise, is presented.

2.4.1 Disturbances and PES

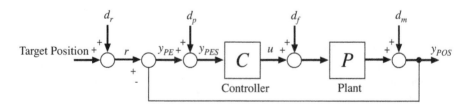

FIGURE 2.17: Block diagram of head-positioning control system.

The feedback control block diagram of the head-positioning control system in HDDs is shown in Figure 2.17, where C and P are controller and plant blocks, y_{PE} is the true positioning error, y_{PES} is the PES, u is the control output or manipulated variable for the plant measured as force or torque, and y_{POS} is the head position as plant output or controlled variable measured as position or angular position. The definitions of the disturbances d_r, d_p, d_f, and d_m are described below.

In HDD head-positioning systems, the target position is generally constant during steady-state for data R/W operations to take place and is usually described in number of *tracks* recorded on the disk. d_r represents the different kinds of disk vibrations and Repeatable Run-Out (RRO) which cause the magnetic head to fluctuate around the target position on the disk. d_p is the demodulation and sensing noises of the PES. d_f shows disturbances acting on the plant as force and torque which are induced by air flow, quantization noise of the D/A converter, and VCM amplifier noise, etc., and d_m is the mechanical vibrations which are uncorrelated to controller output u. y_{PES} is the position error of the head and is measurable, but y_{POS}, the head position and y_{PE}, the true position error (which does not include the sensing noise d_p and other disturbances), are unmeasurable. These concepts are important for the analysis described later.

The target position is generally constant at steady-state. As such, it can be set at zero using a coordinate transform to the origin at the target position, and transfer functions from disturbances to the true positioning error y_{PE} and the head PES y_{PES} can be described by

$$y_{PE} = \frac{1}{1+PC}d_r - \frac{PC}{1+PC}d_p - \frac{P}{1+PC}d_f - \frac{1}{1+PC}d_m, \quad (2.14)$$

$$y_{PES} = \frac{1}{1+PC}d_r + \frac{1}{1+PC}d_p - \frac{P}{1+PC}d_f - \frac{1}{1+PC}d_m. \quad (2.15)$$

The above transfer functions can be simplified by introducing S, T, and Q as

$$S = \frac{1}{1+PC}, \quad (2.16)$$

$$T = \frac{PC}{1+PC} = 1 - S, \quad (2.17)$$

$$Q = \frac{P}{1+PC} = PS. \quad (2.18)$$

The simplified transfer functions are given by

$$\begin{aligned} y_{PE} &= S(d_r - d_m) - Td_p - Qd_f \\ &= S(d_r + d_p - d_m) - Qd_f - d_p, \quad (2.19) \\ y_{PES} &= S(d_r + d_p - d_m) - Qd_f. \quad (2.20) \end{aligned}$$

From (2.19) and (2.20), it can be seen that the transfer function S represents the transfer characteristics from disturbances d_r, d_p, and d_m (which are measured in position) to y_{PE} and y_{PES}, and the transfer function Q represents the transfer characteristics from disturbance d_f (measured in force or torque) to y_{PE} and y_{PES}. It is also clear that the two transfer functions S and Q as well as disturbances determine y_{PE} and y_{PES} uniquely. This is a *direct* problem where one can compute y_{PE} and y_{PES} from the two transfer functions and disturbances from the right hand sides of (2.19) and (2.20).

On the other hand, PES decomposition can be considered as a type of *inverse* problem where y_{PES} on the left hand side of (2.20) can be computed offline from the information of S, Q, and disturbances. In general, the transfer function S in (2.16) is known as the *sensitivity transfer function* and is one of the performance indices of closed-loop systems. It is obvious from (2.17) and (2.18) that the transfer function T is termed the *complementary sensitivity* or *co-sensitivity transfer function*, and Q is determined by S since plant P is fixed and cannot be changed. As such, the sensitivity function S is the most important function which characterizes the positioning error.

2.4.2 Decomposition of Steady-State PES

From the PES, RRO and Non-Repeatable Run-Out (NRRO) can be obtained. NRRO can be further decomposed into mechanical vibrations, PES noise, and torque noise. In this section, these concepts will be illustrated and a basic model for the Track Mis-Registration (TMR) budget design will be presented.

2.4.2.1 RRO and NRRO

The head-positioning error of HDDs can be divided into RRO and NRRO, which are synchronous and asynchronous components with disk rotation, respectively. RRO and NRRO have different causes and effects on the data integrity in HDDs. As such, the evaluation and analysis for RRO and NRRO are carried out separately.

The first step for PES decomposition is to separate PES into RRO and NRRO. RRO can be obtained and removed from PES by averaging it at each servo sector. Because NRRO is asynchronous with disk rotation, it can be reduced by the factor of $\frac{1}{\sqrt{N}}$ after averaging PES in N revolutions. If N is large enough, RRO also shows up as a residual component.

An example of PES data with three thousand revolutions at a revolution time of 6 ms is shown in Figure 2.18. In Figure 2.18, the center line represents RRO which is obtained by the average of measured PES signal at each servo sector. The upper and lower lines show the maximum and minimum values of measured PES at each servo sector, and the width between these two lines indicates the *NRRO peak-to-peak amplitude*, which is roughly equivalent to 6σ of NRRO in most cases.

The main causes of RRO are the repeatable error components between the head trajectory and the target position track on the disk, e.g., disk eccentricity,

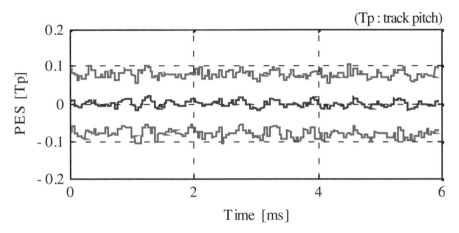

FIGURE 2.18: Time trace of PES.

cogging torque of the spindle motor, media defect noise, and servo sector mislocation, etc., which are written-in during the STW process. These effects can be modeled as d_r added to the target position, as shown in Figure 2.17, and RRO is Sd_r when observed in PES, as shown in (2.20).

The RRO analysis and/or decomposition into each cause is performed in the frequency domain in general. The frequency spectrum of RRO is shown in Figure 2.19. For example, the characteristics of components caused by spindle-

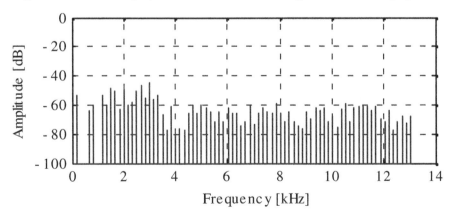

FIGURE 2.19: RRO spectrum.

motor cogging torque are basically determined by the number of slots and poles of the spindle motor. The servo sector mislocation is a kind of written-in NRRO during the STW process. If the head vibrates during the STW process, vibration is recorded on a disk when the head is writing servo sectors, which causes the mislocation. As such, NRRO analysis of the STW process can reveal some of these causes. If the rotational speeds of the drive during normal operations and the STW process are different, the effect of frequency shifts should also be considered in the analysis.

According to (2.20), the frequency spectrum of the main causes of RRO (d_r) can be estimated by applying the gain of S^{-1} to the RRO frequency spectrum. The knowledge of disturbances d_r can help in future analysis.

2.4.2.2   Frequency Spectrum of NRRO

NRRO can be obtained by subtracting the RRO from the PES at each servo sector in the time domain. The frequency spectrum characteristics of NRRO are determined by the two transfer functions S and Q as well as disturbances as depicted in (2.20).

Mechanical vibrations d_m are represented by their respective natural frequencies, while position noise d_p and torque noise d_f are wide band and can be approximated as white noise in most cases. As such, Sd_p and Qd_f in (2.20) should have similar spectra as S and Q, respectively. Even if the position noise d_p and torque noise d_f are colored, the baseline NRRO spectrum should be similar to the shape of $k_{dp}S + k_{df}Q$ as the contributed frequency ranges of S and Q are different, i.e., Q has a large gain in the low frequency range and a small gain in the high frequency range, while S has a small gain in the low frequency range and is about unity gain in the high frequency range, as shown in Figure 2.20. k_{dp} and k_{df} are the amplitude levels of the position noise d_p and torque noise d_f, respectively. Figures 2.20 and 2.21 show an example of

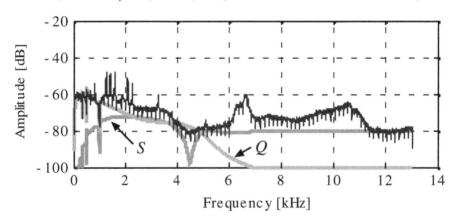

FIGURE 2.20: NRRO.

the frequency spectrum of NRRO which is divided from PES shown earlier in Figure 2.18. The two transfer functions S and Q are fitted to the NRRO baseline and plotted. Estimation of noise level k_{dp} and k_{df} is described in the following section.

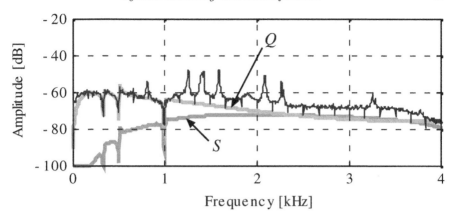

FIGURE 2.21: NRRO up to 4 kHz.

2.4.2.3 Decomposition of NRRO

The NRRO decomposition method consists of the following steps:

1. Fit the transfer function Q with the NRRO baseline spectrum in the low frequency range. In Figures 2.20 and 2.21, Q is fitted with the baseline with emphasis on below 500 Hz where there is no mechanical vibration and the gain of S is relatively much smaller than that of Q. This means that the contribution of Qd_f is largest in this range as compared to other components. Fitting Q with the NRRO baseline spectrum is the same as estimating the amplitude level of the noise d_f;

2. Fit the transfer function S with the NRRO baseline spectrum in the high frequency range. In Figures 2.20 and 2.21, S is fitted with the baseline focusing on above the 12 kHz range where there is no mechanical vibration and the gain of S is relatively much larger than that of Q. Fitting S with the baseline NRRO spectrum estimates the amplitude level of the noise d_p;

3. Merge the fitted lines of Q and S by summing their squared spectra followed by the square root of the merged squared spectra at each frequency point. The result is the total noise baseline, as shown in Figure 2.22;

4. Remove the NRRO components caused by mechanical vibrations by subtracting the total baseline noise from the NRRO spectrum, subtracting the baseline of squared total noise from the squared NRRO spectrum, and taking the square root of the result at each frequency point. This is because the NRRO components caused by the mechanical vibrations result in differences between the total baseline noise and the NRRO spectrum. In many cases, the information of mechanical mode frequency is

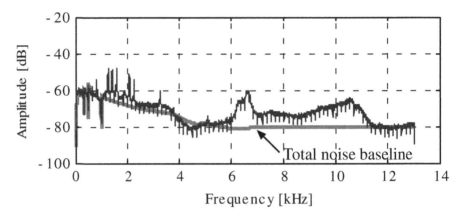

FIGURE 2.22: Baseline of total noise.

available in advance, and this helps in categorizing and removing the components from the NRRO frequency spectrum. An example is shown in Figure 2.23. The vibration components in the range from 1 kHz to

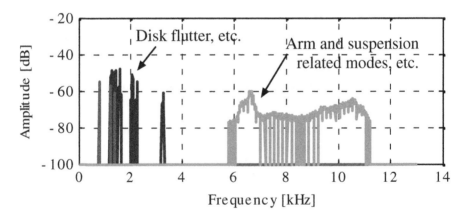

FIGURE 2.23: Mechanical vibrations in NRRO.

3 kHz are the disk flutter modes, and the spectrum in the range above 5 kHz is mainly the carriage-arm and suspension modes. These vibration modes are known as Flow-Induced Vibration (FIV) modes which are excited by air flow, and they are the main detractors of achievable head-positioning accuracies in HDDs [14]; and

5. Decompose the squared residual spectrum into position and torque noise components by dividing according to the relation between the squared S

and Q lines and taking the square root of the results at each frequency point. Figures 2.24 and 2.25 show an example.

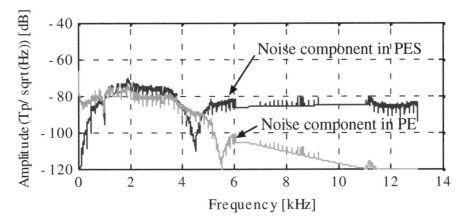

FIGURE 2.24: PES noise in NRRO.

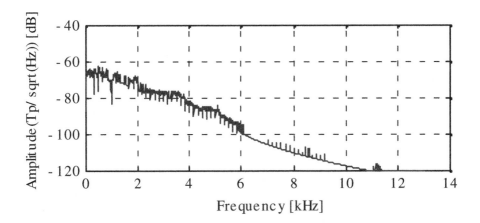

FIGURE 2.25: Torque noise in NRRO.

By comparing (2.14) and (2.15), it can be easily seen that the transfer functions from the position noise d_p to the true positioning error y_{PE} and head PES y_{PES} are different. The true positioning error y_{PE} is not measurable, but can be estimated from y_{PES} by compensating for this difference. This can be achieved by multiplying the position noise component spectrum with the gain PC of the open-loop transfer function. The results are shown in Figure 2.24. In the low frequency range, e.g., less than about 1 kHz in Figure 2.24, the servo loop suppresses any vibrations and noises and hence the

TABLE 2.4: Example of NRRO Decomposition.

		Total	Vibrations	Units
NRRO in PES	σ	2.26×10^{-2}	1.70×10^{-2}	TP
(measured signal)	σ^2 ratio		56.6	%
NRRO in PE	σ	2.20×10^{-2}	1.70×10^{-2}	TP
(true error)	σ^2 ratio		59.8	%
		PES Noise	Torque Noise	Units
NRRO in PES	σ	0.70×10^{-2}	1.31×10^{-2}	TP
(measured signal)	σ^2 ratio	9.6	33.8	%
NRRO in PE	σ	0.46×10^{-2}	1.31×10^{-2}	TP
(true error)	σ^2 ratio	4.4	35.8	%

position-noise component in y_{PES} is small. However, it is large in the true positioning error y_{PE} because the head follows the noise. On the other hand, the opposite happens in the high frequency range. The servo loop cannot suppress the noise since the head does not respond to the noise, and hence the noise component is large in y_{PES} but small in y_{PE}. Table 2.4 shows the results of NRRO decomposition where TP is Track Pitch. In this case, mechanical vibrations are the major components in NRRO followed by torque noise d_f. The contributions of the various disturbances become clear, as shown in Table 2.4, and it becomes easy to calculate the improvements on the PES y_{PES} and true positioning error y_{PE} assuming the same servo loop transfer functions.

Dividing each component by the gain of the corresponding transfer functions in (2.19) and (2.20), the frequency spectrum of each disturbance source can be estimated. The estimated disturbance source can be used for optimizing the transfer function S by changing controller C so that y_{PES} and/or y_{PE} can be minimized.

The NRRO data in this section was measured using a fluid dynamic bearing HDD. As such, there were no mechanical vibration modes and their effects appeared as line spectra. The details on the decomposition of the ball bearing spindle modes are described in [6].

2.4.3 Decomposition of Transient Response

In the previous subsection, PES decomposition is carried out in the steady-state where the target position is kept constant. The root causes for non-zero PES are *stationary* disturbances due to the constant reference or target position and a designed stable servo control system. Decomposition and analysis are also performed in the frequency domain by transforming time-domain PES into its frequency spectrum using the Discrete Fourier Transform (DFT).

In the head-positioning control systems of HDDs, the settling responses are the transients arising from the change from track-seeking mode to track-following mode as the magnetic read head reaches the target track. In order to support the growth in areal storage density and access performance improvements while maintaining the same form factor, the settling responses should be made faster and more accurate. This is getting harder as TPI continues to increase.

In this section, a settling-response analysis method which decomposes the transient response into its contributing residual vibration modes is described. Vibration modes can be excited during the seeking motion when a large acceleration or jerk for high-speed actuation of more than a few thousand tracks has occurred. At the end of the seeking motion, the settling response is typically expected to reduce any type of vibrations to within one-tenth of the track width. Because of this large dynamic range, the mechanical resonant modes which are not captured in the transfer functions may cause residual vibrations. For this application, the DFT is not the best tool for analyzing the settling time response. Instead, the *Prony* experimental modal analysis method is preferred, and it identifies the frequencies, damping ratios, and initial amplitudes of the modes which contribute to the transient response.

A general transient response can be expressed by

$$y(t) = \sum_{r=1}^{n}(a_r e^{s_r t} + a_r^* e^{s_r^* t}), \tag{2.21}$$

$$a_r = U_r + jV_r, \tag{2.22}$$

$$s_r = -\sigma_r + j\omega_{dr}, \tag{2.23}$$

$$\omega_{dr} = 2\pi f_{dr}, \tag{2.24}$$

where a_r is the complex amplitude, $*$ indicates the conjugate of the complex number, V_r and U_r are real and imaginary parts of a_r, respectively. s_r is the complex natural frequency, σ_r is the decay rate, ω_{dr} is the damped natural angular frequency, and f_{dr} is the damped natural frequency. The Prony method aims to identify these modal parameters.

Using Euler's formula, a_r and $y(t)$ can be rewritten as

$$a_r = \sqrt{U_r^2 + V_r^2}e^{j\phi_{0r}}, \tag{2.25}$$

$$\phi_{0r} = tan^{-1}\left(\frac{V_r}{U_r}\right), \tag{2.26}$$

$$y(t) = \sum_{r=1}^{n}\{a_{0r}e^{-\sigma_r t}cos(\omega_{dr}t + \phi_{0r})\}, \tag{2.27}$$

$$a_{0r} = 2\sqrt{U_r^2 + V_r^2}, \tag{2.28}$$

$$\omega_{dr} = \omega_{nr}\sqrt{1 - \zeta_r^2}, \tag{2.29}$$

$$\sigma_r = \zeta_r\omega_{nr}, \tag{2.30}$$

$$\zeta_r = \frac{\frac{\sigma_r}{\omega_{dr}}}{\sqrt{1 + \left(\frac{\sigma_r}{\omega_{dr}}\right)^2}}, \tag{2.31}$$

where ϕ_{0r} is the initial phase, a_{0r} is the initial amplitude, ζ_r is the damping ratio, and ω_{nr} is the undamped natural angular frequency. In the case where $V_r = 0$ and $\omega_{dr} = 0$, the mode is known as an *overdamped* mode.

The closed-loop transfer function of stable servo systems can be described in general using (2.32) as [16]

$$Y(s) = \frac{N(s)}{\prod_i (s + \alpha_i) \prod_k (s^2 + 2\zeta_k \omega_{nk} s + \omega_{nk}^2)}, \tag{2.32}$$

where α_i is the natural angular frequency of the first-order delay element, ζ_k and ω_{nk} are the damping ratio and angular natural frequency of the second-order delay element, respectively. If $Y(s)$ has no multiple roots, the step response of the servo system depicted in (2.32) can be expressed according to the expansion theorems as

$$y(t) = Y(0) + \sum_i A_i \exp^{-\alpha_i t} + \sum_k B_k \exp^{-\zeta_k \omega_{nk} t} \cos\left(\omega_{nk}\sqrt{1 - \zeta_k^2}t + \phi_k - \frac{\pi}{2}\right), \tag{2.33}$$

where $Y(0)$ is the stationary error. The servo system compensates the disturbances so that $Y(0)$ becomes zero in most cases. Besides the second terms which are the responses of overdamped modes, the third terms are basically the same as (2.21) and (2.27). This shows that the Prony method can be applied to the analysis of settling responses in HDDs, and more details on the case which contains overdamped modes represented by the second term in (2.33) can be found in [15].

Examples of the settling response decomposition using the Prony method in experiments are shown in Figures 2.26, 2.27, 2.28, 2.29, and 2.30. An example of a seek time response is shown in Figure 2.26 where the upper plot is the VCM current and the lower plot is the corresponding PES. The settling response was measured using the PES as shown. In this case, the point when the time of the settling response is zero is set approximately to be the time when the overdamped modes corresponding to the second term in (2.33) become small enough. Reducing NRRO and RRO by averaging the seek responses under the same conditions to isolate the settling response is also a critical step before starting the analysis.

The averaged result of one hundred measurements is shown in Figure 2.26. In Figure 2.26, the VCM current during settling and track-following is much smaller than that during seeking. Hence, it can be considered that the residual vibrations are excited during track-seeking and appear in the settling responses as a result.

The identification results of the modes using the Prony method for damping ratio of less than 0.4 are shown in Figure 2.27. The time responses of the

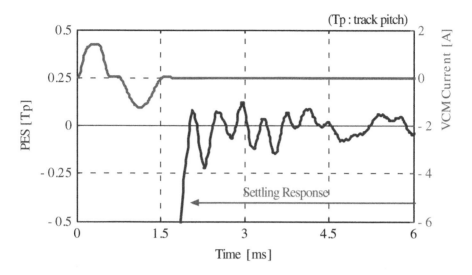

FIGURE 2.26: Example of settling response.

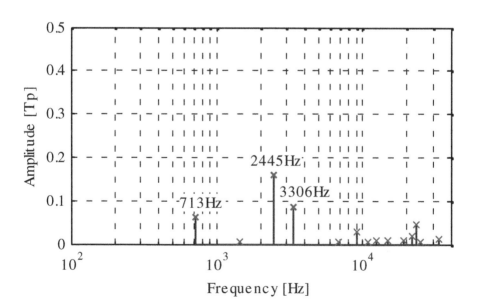

FIGURE 2.27: Residual modes in settling response.

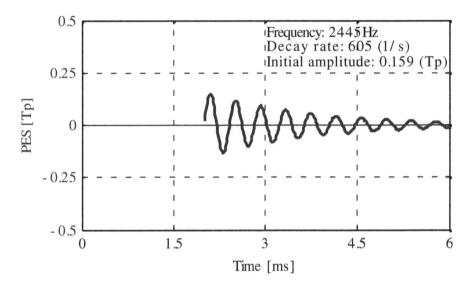

FIGURE 2.28: Response of mode at 2445 Hz.

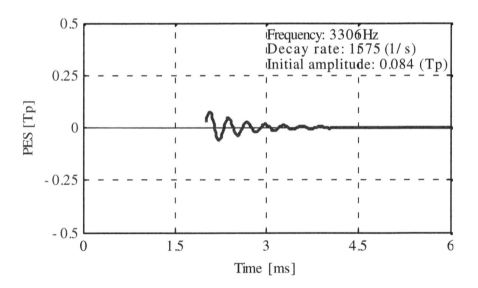

FIGURE 2.29: Response of mode at 3306 Hz.

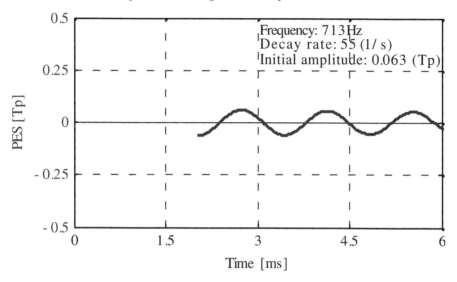

FIGURE 2.30: Response of mode at 713 Hz.

three major modes at 2445 Hz, 3306 Hz, and 713 Hz are shown in Figures 2.28, 2.29, and 2.30, respectively. The responses are simulated using the identified modal parameters. The 2445 Hz mode has a smaller decay rate as compared to the 3306 Hz mode and hence the residual vibrations last longer. The initial amplitude of the 713 Hz mode is smaller as compared to the other two modes, but its decay rate is smallest and its amplitude becomes the biggest after some time. The decay rate (which is not considered in the steady-state positioning error analysis) is one of the most important indices besides amplitude during transient response analysis. It should be noted that the modal decay rate is determined by both the mechanical characteristics and servo loop transfer functions.

Measurement noise, e.g., NRRO and RRO, as well as model reduction and parameter setting for modal identification, e.g., modal number, are key considerations to be taken into account if a good settling response decomposition using the Prony method is required. More details and guidelines can be found in [7][8][15]. Down-sampling the measurements and pre-filtering for identification of the low frequency modes are also other effective techniques from an engineering point of view.

Bibliography

[1] C. D. Mee and E. D. Daniel, *Magnetic Recording Volume 2: Computer Data Storage*, Chapter 2, McGraw-Hill, 1988.

[2] G. Franklin, J. D. Powell, and M. L. Workman, *Digital Control of Dynamic Systems*, 2nd Edition, Addison Wesley, 1990.

[3] *Dynamics and Control of Information Equipment*, Yokendo Ltd., 1996 (in Japanese).

[4] T. Yamaguchi, "Nano-Mechatronics Technologies for Head Positioning System of Hard Disk Drives," in *Proceedings of the Conference on Information, Intelligence and Precision Equipment, Japan Society of Mechanical Engineers*, No. 01-8, pp. 136–141, 2001 (in Japanese).

[5] Y. Takahashi, M. J. Rabins, and D. M. Auslander, *Control and Dynamics Systems*, Addison-Wesley Publishing Company, 1972.

[6] H. Numasato, K. Usui, Y. Hamada, and T. Yamaguchi, "An Analysis Method of Head Positioning Error for Magnetic Disk Drives (An Analysis Method of PES in Track Following Mode)," *Transactions of the Japan Society of Mechanical Engineers: Series C*, Vol. 65, No. 636, pp. 3245–3251, August 1999 (in Japanese).

[7] A. Nagamatsu, *Introduction to Modal Analysis*, Corona Publishing Co. Ltd., 1993 (in Japanese).

[8] A. Nagamatsu, *Modal Analysis*, Baifukan Co. Ltd., 1985 (in Japanese).

[9] D. Abramovitch, F. Wang, and G. F. Franklin, "Disk Drive Pivot Nonlinearity Modeling. I: Frequency Domain," in *Proceedings of the American Control Conference*, Vol. 3, pp. 2600–2603, 1994.

[10] T. Atsumi, "Head-Positioning Control Using Virtual Resonant Modes in a Hard Disk Drive," *Transactions of the Japan Society of Mechanical Engineers: Series C*, Vol. 71, No. 706, pp. 1914–1919, 2005 (in Japanese).

[11] K. J. Åstrom and B. Wittenmark, *Computer-Controlled Systems*, Prentice-Hall, 1984.

[12] T. Atsumi, A. Arisaka, T. Shimizu, and T. Yamaguchi, "Vibration Servo Control Design for Mechanical Resonant Modes of Hard-Disk-Drive Actuator," *JSME International Journal: Series C*, Vol. 46, No. 3, pp. 819–827, 2003.

[13] T. Yamaguchi and T. Atsumi, "Issues on Vibration Control and Settling Control of Hard Disk Drive Servo System," in *Proceedings of the 33rd SICE Symposium on Control Theory*, pp. 421–426, 2004 (in Japanese).

[14] S. Imai and H. Numasato, "Estimation Method of Head-Positioning Error due to Disk Flutter," *Transactions of the Japan Society of Mechanical Engineers: Series C*, Vol. 65, No. 636, pp. 3075–3081, August 1999 (in Japanese).

[15] H. Numasato, K. Ito, and S. Saegusa, "An Analysis Method of Head Positioning Error for Magnetic Disk Drives (Settling Response Analysis)," *Transactions of the Japan Society of Mechanical Engineers: Series C*, Vol. 65, No. 638, pp. 4093–4099, October 1999 (in Japanese).

[16] T. Suzuki, *Exercise on Theory of Automatic Control*, Gakuken Co., 1984 (in Japanese).

[17] D. Abramovitch, T. Hurst, and D. Henze, "An Overview of the PES Pareto Method for Decomposing Baseline Noise Sources in Hard Disk Position Error Signals," *IEEE Transactions on Magnetics*, Vol. 34, No.1, pp. 17–23, 1998.

Chapter 3

Basic Approach to High-Speed Precision Motion Control

Atsushi Okuyama

Tokai University

Takashi Yamaguchi

Ricoh Company Ltd.

Takeyori Hara

Toshiba Corp.

Mitsuo Hirata

Utsunomiya University

3.1   Introduction to Mode Switching Control (MSC)

The block diagram for a basic servo control system in Hard Disk Drives
(HDDs) is shown in Figure 3.1.

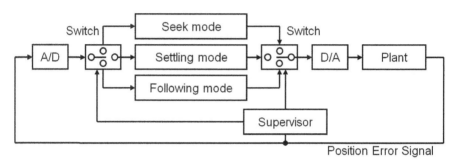

FIGURE 3.1: A block diagram of a servo control system in an HDD.

The main purpose of servo control in HDDs is to move the magnetic R/W
head or head assembly with a high speed to the target position as quickly
as possible, and for the head to follow the target position as precisely as
possible. To achieve this, optimal controllers are usually designed for high-
speed motion and/or high-precision positioning in HDDs. Based on the desired
mode of operation, the corresponding controller is then chosen and this scheme
is known as *Mode Switching Control* (MSC). The basic idea of MSC is to
switch the control mode based on a pre-defined *criterion* using *measured* or
estimated state variables. The MSC structure has been widely applied in many
industries, and this idea is related to the *switched systems* theory commonly
encountered in *hybrid systems* and is also related to *robust control* and *adaptive
control* theories.

In HDDs, the MSC switches the head-positioning control system among
the following modes and objectives:

1. *Track-seeking mode*: trajectory planning and design of reference trajectory, as well as tracking servo control design for the reference;

2. *Track-settling mode*: end-point control to handle residual vibrations and
   large initial states, etc.;

3. *Track-following mode*: trajectory following and design of servo control
   system, focusing mainly on appropriate compromise between attaining
   disturbance suppression capability and robust stability;

for enhanced vibration suppression and rejection. It is important to understand the operations of these servo control methods. In practice, the track-settling mode is essential in order to reduce the transient response after mode switching. The details for all three operating modes are described in this chapter.

3.2 Track-Seeking: Fast Access Servo Control

In this section, the Two-Degree-of-Freedom (TDOF) and access servo control considering saturation techniques are presented to improve the track-seeking performance of HDDs.

3.2.1 Two-Degrees-of-Freedom (TDOF) Control

In this section, a feedforward type TDOF control is presented. This section explains the advantages and structure of TDOF control, as well as concepts on Zero-Phase Error Tracking Control (ZPETC) and reference trajectory design that are crucial in the design of a feedforward type TDOF control.

3.2.1.1 Advantages of TDOF Control

Several closed-loop transfer functions, such as transfer function from the reference trajectory to the controlled variable, can be determined from a feedback control system. The degrees of freedom of a control system is defined as the number of closed-loop transfer functions that can be selected *independently* [1].

In the case of the unity feedback control system which has been widely used as the head-positioning control system of an HDD, as shown in Figure 3.2, the closed-loop transfer functions are

$$H_{yr} = \frac{PC_1}{1 + PC_1} := T, \tag{3.1}$$

$$H_{yw} = \frac{P}{1 + PC_1}, \tag{3.2}$$

where H_{yr} is the transfer function from the reference r to the controlled variable y, and H_{yw} is the transfer function from the disturbance w to the controlled variable y. If either transfer function is designed by using the feedback controller C_1, the other transfer function is uniquely decided. Therefore, the degrees of freedom of the control system is as shown in Figure 3.2. Such a control system is called a One-Degree-of-Freedom (ODOF) control system.

In the case of the control system shown in Figure 3.3, a feedforward controller C_2 is added to the control system in Figure 3.2. The closed-loop transfer function from the reference r to the controlled variable y is

$$H_{yr} = TC_2. \tag{3.3}$$

The transfer function from the disturbance w to the controlled variable y is (3.2). Even if H_{yw} is designed by using the feedback controller C_1, H_{yr} is independently designed by using the feedforward controller C_2. Therefore, the degrees of freedom of the control system is two, as shown in Figure 3.3, and such a control system is called a Two-Degrees-of-Freedom (TDOF) control system.

In the head-positioning control system of HDDs, the required performances are precise positioning and fast access. This is not easy for the ODOF control system. For the case where the power amplifier is not saturated during track-seeking, which usually happens for short-span seeking, the TDOF control is already a very popular control method. On the other hand, for the case where the power amplifier is saturated during track-seeking, which usually happens for long-span seeking, the access servo control considering saturation is applied.

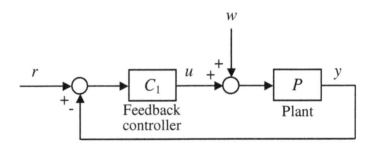

FIGURE 3.2: Unity feedback ODOF control system.

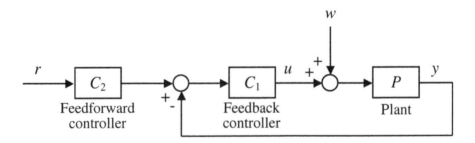

FIGURE 3.3: Filter-type expression of TDOF control system.

3.2.1.2   Structure of TDOF Control

Various TDOF control structures have been proposed and detailed analyses including equivalent transformation have been studied [2]. This section focuses on the feedforward type expression of the TDOF control system, as shown in Figure 3.4. The closed-loop transfer function from the reference r to the controlled variable y is

$$H_{yr} = \frac{PC_2 + PC_1}{1 + PC_1}. \tag{3.4}$$

The transfer function from the disturbance w to the controlled variable y is described by (3.2). Therefore, H_{yr} and H_{yw} can be designed independently by using C_2 and C_1, respectively. In particular, when the feedforward controller C_2 is designed as

$$C_2 = P^{-1}, \tag{3.5}$$

the ideal closed-loop transfer function from the reference r to the controlled variable y is achieved. In order words, $H_{yr} = 1$ is realized.

The feedforward controller C_2 in (3.5) is the inverse dynamics of the plant P. If the plant P has unstable zeros, the feedforward controller C_2 becomes unstable. Such an unstable feedforward controller C_2 is unusable. Thus, the design of the feedforward controller C_2 involves determining the stable inverse dynamics from a plant with unstable zeros. From this viewpoint, the ZPETC methodology described in the following section has been proposed for discrete-time plant with unstable zeros [3]–[8].

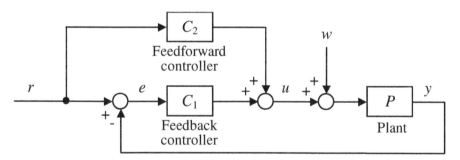

FIGURE 3.4: Feedforward type expression of TDOF control system.

3.2.1.3   Zero-Phase Error Tracking Control (ZPETC)

Let the discrete time transfer function for the closed-loop system be expressed as

$$G[z^{-1}] = \frac{z^{-d}B[z^{-1}]}{A[z^{-1}]}, \tag{3.6}$$

where z^{-1} is the one sampling delay, z^{-d} is the d-step delay normally caused by the delay in the plant and computation, and

$$B[z^{-1}] = b_0 + b_1 z^{-1} + \cdots + b_m z^{-m}, \quad b_0 \neq 0, \tag{3.7}$$

$$A[z^{-1}] = 1 + a_1 z^{-1} + \cdots + a_n z^{-n}. \tag{3.8}$$

In order to deal with the unstable zeros, $B[z^{-1}]$ is divided into two parts with

$$B[z^{-1}] = B^-[z^{-1}]B^+[z^{-1}], \tag{3.9}$$

where $B^-[z^{-1}]$ is the sth degree monic polynomial with unstable zeros which exist on or outside the unit circle in the z-plane, and $B^+[z^{-1}]$ is the $(m-s)^{\text{th}}$ degree polynomial with stable zeros. The inverse dynamics (which is the feedforward controller designed by the ZPETC theory) can be determined as

$$G_{ZPETC}[z^{-1}] = \frac{A[z^{-1}]B^{-*}[z^{-1}]}{B^+[z^{-1}](B^-[1])^2}, \tag{3.10}$$

where

$$B^{-*}[z^{-1}] = z^{-s}B^-[z]. \tag{3.11}$$

By placing the feedforward controller described by (3.10) in front of the transfer function in (3.6), the transfer function from the reference r to the controlled variable y can be obtained as

$$\begin{aligned}
y[k] &= \frac{z^{-d}B[z^{-1}]}{A[z^{-1}]} \frac{A[z^{-1}]B^{-*}[z^{-1}]}{B^+[z^{-1}](B^-[1])^2} r[k+s+d] \\
&= \frac{B^-[z^{-1}]B^{-*}[z^{-1}]}{(B^-[1])^2} r[k+s] \\
&= \frac{B^-[z^{-1}]B^-[z]}{(B^-[1])^2} r[k],
\end{aligned} \tag{3.12}$$

where $r[k+s+d]$ is the $(s+d)$ step ahead reference trajectory. The feedforward controller designed by the ZPETC theory utilizes the future values to compensate for the delay in the closed-loop transfer function, and cancels the poles and cancellable zeros of the closed-loop transfer function. The following relation can be easily verified by direct substitution of $z = exp(j\omega T_s)$

$$\text{Im} \left\{ B^-(e^{-j\omega T_s})B^-(e^{j\omega T_s})/B^-(1)^2 \right\} = 0, \quad 0 \leq \omega \leq \pi/T_s, \tag{3.13}$$

where T_s is the sampling time. (3.13) implies that the phase error between $r[k]$ and $y[k]$ is zero for all frequencies. However, the gain error between $r[k]$ and $y[k]$ remains. The gain between $r[k]$ and $y[k]$ is close to one at low frequencies and decreases at high frequencies.

As an example, the controlled plant is assumed to be a pure double integrator and is depicted by

$$G_c(s) = \frac{k_c}{s^2}, \tag{3.14}$$

where k_c is the gain. The discrete-time plant which includes the characteristics of a Zero-Order Hold (ZOH) and sampler can be obtained as

$$G[z^{-1}] = k\frac{z^{-1}(1 + z^{-1})}{(1 - z^{-1})^2}, \quad \because k := k_c\frac{T_s^2}{2}. \tag{3.15}$$

The numerator is divided into two parts with

$$B^-[z^{-1}] = 1 + z^{-1}, \tag{3.16}$$

$$B^+[z^{-1}] = 1. \tag{3.17}$$

Thus, (3.11) is

$$B^{-*}[z^{-1}] = z^{-1}B^-[z] = z^{-1}(1 + z) = 1 + z^{-1}. \tag{3.18}$$

The feedforward controller is determined as

$$
\begin{aligned}
G_{ZPETC}[z^{-1}] &= \frac{A[z^{-1}]B^{-*}[z^{-1}]}{B^+[z^{-1}](B^-[1])^2} \\
&= \frac{(1 - z^{-1})^2}{k}\frac{1 + z^{-1}}{4}.
\end{aligned} \tag{3.19}
$$

The transfer function from the reference r to the controlled variable y becomes

$$
\begin{aligned}
y[k] &= k\frac{z^{-1}(1 + z^{-1})}{(1 - z^{-1})^2}\frac{(1 - z^{-1})^2}{k}\frac{1 + z^{-1}}{4}r[k + 2] \\
&= \frac{(1 + z)(1 + z^{-1})}{4}r[k].
\end{aligned} \tag{3.20}
$$

The frequency responses for the conditions of the benchmark problem are plotted in Figures. 3.5–3.8, where Figure 3.5 is a plot for the discrete time plant of (3.15) and Figure 3.6 is a plot for the inverse model of (3.19). Figure 3.7 is a plot for the inverse model with two look ahead steps $z^2G_{ZPETC}[z^{-1}]$, and Figure 3.8 is a plot for the transfer function from the reference r to the controlled variable y. The phase shift is zero for all frequencies, while the gain decreases at high frequencies. Such gain error degrades the tracking performance. A lot of methods have been proposed to compensate the gain error of the ZPETC. However, these methods cannot compensate the gain error completely. The ZPETC is basically a design method for single-rate systems. The Perfect Tracking Control (PTC) based on multi-rate sampling technique has been proposed for multi-rate systems. The feedforward controller designed by the PTC theory can cancel both phase and gain errors.

3.2.1.4   Reference Trajectory

Even if the feedforward controller can achieve an ideal characteristic where both phase and gain error becomes zero, the tracking performance of the

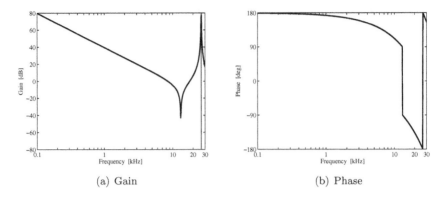

(a) Gain   (b) Phase

FIGURE 3.5: Frequency response of plant.

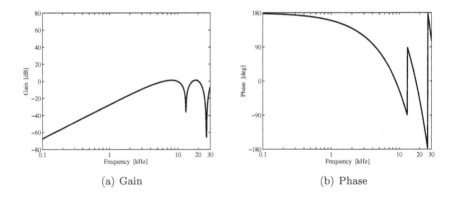

(a) Gain   (b) Phase

FIGURE 3.6: Frequency response of inverse model.

TDOF control system depends on the given reference trajectory. Therefore, a lot of design methods for generating the reference trajectory have been proposed. Optimal trajectory design methods that minimize a given performance index such as time, power, and jerk have been proposed in [9]–[13].

As an example, the minimum jerk trajectory is explained. This trajectory minimizes the performance index that involves the integrated square of jerk. This trajectory design is based on a pure double-integrator system. Figure 3.9 shows the augmented system where x_1 is the position, x_2 is the velocity, x_3 is the control input that corresponds to the acceleration, and v is the derivative of the control input. The performance index is described as

$$J = \int_0^T \left(\frac{du}{dt}\right)^2 dt = \int_0^T v^2 dt. \tag{3.21}$$

The state equation and boundary conditions of the augmented system as

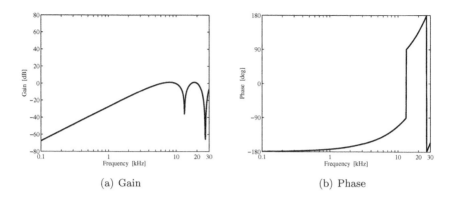

(a) Gain   (b) Phase

FIGURE 3.7: Frequency response of inverse model with two look ahead steps.

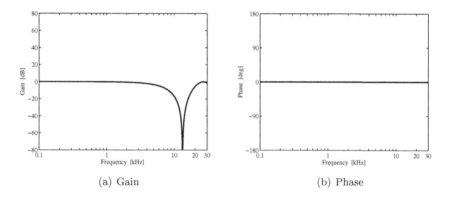

(a) Gain   (b) Phase

FIGURE 3.8: Frequency response from reference trajectory to plant output.

shown in Figure 3.9 are

$$\frac{d}{dt}\begin{bmatrix} x_1 \\ x_2 \\ x_3 \end{bmatrix} = \begin{bmatrix} 0 & 1 & 0 \\ 0 & 0 & 1 \\ 0 & 0 & 0 \end{bmatrix}\begin{bmatrix} x_1 \\ x_2 \\ x_3 \end{bmatrix} + \begin{bmatrix} 0 \\ 0 \\ 1 \end{bmatrix} v$$

$$:= Ax + Bv, \tag{3.22}$$

$$x(0) = \begin{bmatrix} 0 & 0 & 0 \end{bmatrix}^T, \quad x(T) = \begin{bmatrix} L & 0 & 0 \end{bmatrix}^T, \tag{3.23}$$

where T is the specified time for movement and L is the specified distance for movement. To determine the trajectory, the Lagrange multiplier is applied as

$$H = v^2 + \lambda^T(Ax + Bv), \tag{3.24}$$

$$\frac{d\lambda}{dt} = -\frac{\partial H}{\partial x}, \tag{3.25}$$

where H is the Hamiltonian and λ is the Lagrange multiplier. By differentiating H with respect to v, we get

$$\frac{\partial H}{\partial v} = 2v + \lambda^T B = 0, \tag{3.26}$$

$$v = -\frac{1}{2}\lambda^T B. \tag{3.27}$$

As a result, the minimum jerk trajectory can be obtained as

$$x_1(t) = 60L\left\{\frac{1}{10}\left(\frac{t}{T}\right)^5 - \frac{1}{4}\left(\frac{t}{T}\right)^4 + \frac{1}{6}\left(\frac{t}{T}\right)^3\right\}, \tag{3.28}$$

$$x_2(t) = \frac{30L}{T}\left\{\left(\frac{t}{T}\right)^4 - 2\left(\frac{t}{T}\right)^3 + \left(\frac{t}{T}\right)^2\right\}, \tag{3.29}$$

$$x_3(t) = \frac{60L}{T^2}\left\{2\left(\frac{t}{T}\right)^3 - 3\left(\frac{t}{T}\right)^2 + \left(\frac{t}{T}\right)\right\}, \tag{3.30}$$

where the horizontal axis is the normalized time t/T and L is the distance that corresponds to one track. These trajectories are plotted in Figure 3.10 where each trajectory is normalized by each maximum amplitude and T is six times the sampling time.

These optimal trajectory design methods are basically based on the continuous time plant model. Therefore, they cannot compensate for the effects of a ZOH. From this viewpoint, the Final-State Control (FSC) is proposed for a discrete-time plant with ZOH.

FIGURE 3.9: Augmented system.

3.2.2 Access Servo Control Considering Saturation

This section explains the access servo control considering saturation technique, which in essence is a velocity control system using a reference velocity trajectory. To design a smooth reference velocity trajectory that considers saturation and is robust to system uncertainties and measurement noise, the Proximate Time-Optimal Servomechanism (PTOS) methodology is used and is presented.

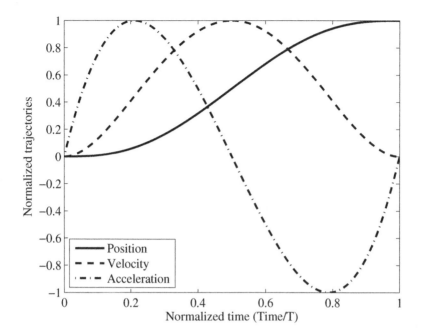

FIGURE 3.10: Minimum jerk trajectory.

3.2.2.1 Basic Structure of Access Servo Control

Bang-bang control has been known as the time-optimal control solution of linear systems with bounded control inputs [14][15]. The control input is restricted to be between a lower and an upper bound. Bang-bang control switches from maximum input to minimum input at certain times. This means that maximum acceleration is applied until the switching point, after which maximum deceleration is applied until the desired position is reached. However, since such a control input excites mechanical resonant modes, bang-bang control is not necessarily suitable for a controlled plant with mechanical resonant modes.

In the access servo control for an HDD, it is ideally desirable to apply maximum acceleration followed by maximum deceleration. However, to prevent mechanical vibrations, the following control method has been widely used, i.e., maximum acceleration is applied up to the limit of the power amplifier where it is switched to deceleration once the head velocity reaches the velocity defined by the deceleration trajectory. Therefore, the access servo control system for an HDD has been generally designed as a *velocity* control system with a reference velocity trajectory [16]–[19].

The basic structure of access servo control for an HDD is shown in Figure 3.11. To ensure that the head settles on a target track, the reference velocity trajectory is calculated according to the current position error in real-time. If the controlled plant is a pure double integrator and the head velocity can be directly detected, the velocity control system can be transformed to that shown in Figure 3.12. In this case, the closed-loop transfer function from the reference velocity r to the head-velocity v is expressed by the following first-order system as

$$v = \frac{K_v K_1 K}{s + K_v K_1 K} r, \tag{3.31}$$

where K is the feedback gain, K_1 is the plant input gain, and K_v is the velocity detector gain. The feedback gain K is designed so that the gain crossover frequency of the velocity control system becomes the specified frequency. Furthermore, as this velocity control system is a Type-I system, steady-state error occurs when a ramp input is applied. To follow a ramp input, the reference acceleration trajectory (which is the derivative of the reference velocity trajectory) is applied to the velocity control system. In this case, the velocity control system becomes the TDOF control system, and the closed-loop transfer function from the reference velocity to the head velocity becomes one, i.e., ideal tracking performance is achieved.

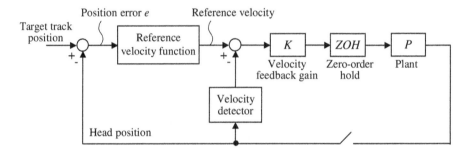

FIGURE 3.11: Basic structure of access servo control for HDD.

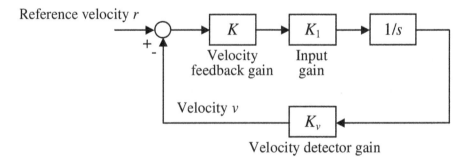

FIGURE 3.12: Block diagram of velocity servo control system.

3.2.2.2 Reference Velocity Trajectory

An example of a reference velocity trajectory is shown in Figure 3.13, where V_{max} is the maximum velocity, A_{max} is the maximum acceleration, L_{max} is the maximum stroke, t_1 is the acceleration period, t_2 is the constant velocity period, t_3 is the deceleration period, and t_{total} is the whole period. The following equations hold

$$t_1 = t_3 = \frac{V_{max}}{A_{max}}, \tag{3.32}$$

$$t_{total} = t_1 + t_2 + t_3, \tag{3.33}$$

$$L_{max} = \frac{1}{2} t_1 V_{max} + t_2 V_{max} + \frac{1}{2} t_3 V_{max} = V_{max} \left(\frac{1}{2} t_1 + t_2 + \frac{1}{2} t_3 \right), \tag{3.34}$$

$$t_2 = \frac{|L_{max}|}{V_{max}} - \frac{|t_1|}{2} - \frac{|t_3|}{2}. \tag{3.35}$$

Both t_1 and t_3 are determined according to (3.32), and t_2 is determined according to (3.35). When t_2 is less than zero, it is necessary to shorten both t_1 and t_3.

The reference velocity trajectory during deceleration can be defined as a function that provides the relationship between position error and velocity. If the acceleration during deceleration is assumed to be A_{dec}, the relationship between velocity v and time t is expressed as

$$v = A_{dec} t. \tag{3.36}$$

The above-mentioned time t progresses in the opposite direction, as shown in Figure 3.13. The position error e at time t is calculated from the area of the shaded portion in Figure 3.13 as

$$e = \frac{1}{2} A_{dec} t^2, \tag{3.37}$$

$$t = \sqrt{\frac{2e}{A_{dec}}}. \tag{3.38}$$

By substituting (3.38) into (3.36), the reference velocity function is obtained as

$$v = \sqrt{2 A_{dec} e}. \tag{3.39}$$

The following function that considers the moving direction has been generally used

$$v = \text{sgn}(e) \sqrt{2 A_{dec} |e| - c^2} - c, \tag{3.40}$$

where c is the adjustment gain. To achieve a smooth deceleration, various reference velocity functions such as an exponential function have been proposed.

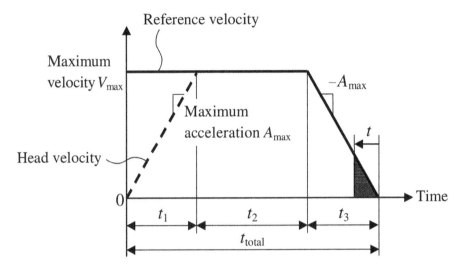

FIGURE 3.13: Example of reference velocity trajectory.

3.2.2.3 Proximate Time-Optimal Servomechanism (PTOS)

Time-optimal control plays an important role in achieving fast point-to-point motion. However, time-optimal control is not robust with respect to system uncertainties and measurement noises. To overcome such a drawback, the PTOS has been proposed [20]–[23].

The basic structure of PTOS is shown in Figure 3.14. The reference velocity function of the PTOS is expressed as

$$v = \begin{cases} \text{sgn}(e)\sqrt{2A_{dec}|e|} - V_{off} & for \quad |e| > e_l, \\ \frac{A_{dec}}{2V_{off}}e & for \quad |e| \le e_l, \end{cases} \tag{3.41}$$

$$e_l = \frac{2V_{off}^2}{A_{dec}}, \tag{3.42}$$

where e_l is the boundary region in which the PTOS operates as a linear control law, and V_{off} is the velocity offset. The reference velocity function of the PTOS consists of a non-linear part and a linear part. When the position error e is greater than the boundary region e_l, the reference velocity trajectory v is generated by the non-linear part of the reference velocity function. This means that the PTOS operates as a non-linear control law. Furthermore, when the position error e becomes so large that saturation occurs, maximum acceleration is applied. Once the position error e becomes small, the inner velocity feedback loop can follow the reference velocity trajectory v. When the position error e is within the boundary region e_l, the reference velocity trajectory v is generated by the linear part of the reference velocity function. This means that the PTOS operates as a linear control law. Since the velocity at switching from the non-linear part to the linear part is in agreement with the velocity offset V_{off}, the control law can be continuously switched from the non-linear control law to the linear control law. PTOS is an excellent control method where both the saturation problem and smooth transition problem have been solved.

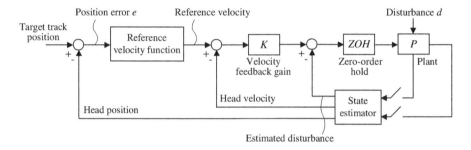

FIGURE 3.14: Basic structure of PTOS.

3.3 Track-Settling: Initial Value Compensation (IVC)

In this section, the Initial Value Compensation (IVC) technique that improves the transient performance after mode-switching is presented. This technique is applied to MSC in HDDs and its effectiveness is verified in simulations as well as experimental implementations in actual HDDs.

3.3.1 Concept of IVC

The logic behind switching for MSC in HDDs is simple. The supervisor switches the controller for high-speed movement to the controller for precision positioning, and the sequence for switching is pre-determined. During switching, the state variables of controllers are discontinuously updated, and switching from one controller to the other brings about undesirable transient responses. As such, the main purpose of controller switching is to compensate for the transient responses after MSC. Several ideas have been proposed in current literature, e.g., setting appropriate initial values for the initial states of the controller after mode switching and designing an optimal trajectory so that the plant state variables such as position, velocity, and acceleration are appropriate during mode switching.

3.3.1.1 Initialization of Controller State Variable

The transient response after mode switching depends on the initial values of the state variables. As such, it is expected that improvements in the transient response can be achieved by setting appropriate initial values for the controller state variables during mode switching. This is made possible because the state variables of the controller can be easily overwritten in the microprocessor or Digital Signal Processor (DSP) program. Currently, there are two methods to calculate the required initial values, namely, the *online calculation* method based on detected plant state variables and the *offline calculation* method.

In the online calculation method, the *Initial Value Compensation* (IVC) method has been proposed to calculate the initial values of the controller using the product of the detected plant state variables and a predefined matrix. In the first version of the proposed IVC which was also applied to actual HDDs, the coefficient matrix minimizing the quadratic equation of the closed-loop state variables was designed [24]. The coefficient matrix which reallocates some zeros of the transfer function from initial states of the plant to the controlled variables on locations of the closed-loop poles such that the poles which give undesirable transient responses can be cancelled by the newly assigned zeros was also designed [25]. Since then, various methods based on the IVC have been proposed, e.g., the IVC design which assigns only the desirable trajectory of transient response by adding a new pole and cancelling all other poles with zeros [26], the IVC design which minimizes the quadratic function of input and output signals of the plant [27], the IVC design for an observer-based state feedback system [28], the IVC design which considers robustness against measurement noise in the state variables of the plant [29], the IVC design which guarantees minimum settling time considering fluctuation of the state variables of the plant during mode switching [30], and the IVC design which considers mechanical resonant modes [31], just to name a few. As setting new initial values for the controller during mode switching is equivalent to *adding a feedforward input* to the output of the controller, several IVC designs calculating the feedforward input vector based on the measured state variables of the plant during mode switching have also been proposed [32]–[34].

As for the latter, an offline IVC design method which minimizes an error system index representing the steady-state of the closed-loop response after mode switching has been proposed [35]. In addition, general and theoretical approaches to the design of IVC have also been proposed [36][37].

3.3.1.2 Design of Mode Switching Condition

The supervisor for switching control controls the time of mode switching and decides on the controller to be switched to. In MSC for HDDs, the controllers and the sequence of switching are predetermined. As such, the main design issue and objective are to decide on the *time* of mode switching. When the IVC design is applied to mode switching, the initial state variables of the controller are calculated based on the state variables of the plant. As such, the number of state variables of the closed-loop system is reduced to the number of state variables of the plant, and the design of the mode switching condition is the problem of finding an *optimal* combination of state variables of the plant such as position, velocity, and acceleration, etc. In practice, it is possible to find a set of optimal initial state variables that minimizes the quadratic function of position or control input [38]. This can be done by designing a reference trajectory which passes through such a set of state variables. An appropriate switching condition will ensure smooth transition of the control input, which will in turn reduce the acoustic noise during mode switching [39].

3.3.2 IVC Design Method

The following conditions are assumed when designing the IVC. The track-following mode is a precise positioning mode and is a stable Single-Input-Single-Output (SISO) system, and the transfer functions of the plant and controller are proper. As the controller is implemented in a microprocessor or DSP, the corresponding equations are obtained in the discrete-time domain.

First, the state-space equations of the controller and the plant are represented by

$$\begin{align}
x_p[k+1] &= A_p x_c[k] + B_p u[k], \tag{3.43}\\
y[k] &= C_p x_p[k], \tag{3.44}
\end{align}$$

and

$$\begin{align}
x_c[k+1] &= A_c x_c[k] + B_c(r[k] - y[k]), \tag{3.45}\\
u[k] &= C_c x_c[k] + D_c(r[k] - y[k]), \tag{3.46}
\end{align}$$

respectively, where x_p is the mth-order state vector of the plant, x_c is the nth-order state vector of the controller, u is the control input to the plant, r is the reference, and y is the controlled variable, i.e., the head position in HDD applications. A_p, B_p, C_p, A_c, B_c, C_c, and D_c are real matrices of dimensions $m \times m$, $m \times 1$, $1 \times m$, $n \times n$, $n \times 1$, $1 \times n$, and 1×1, respectively. The time during mode switching is denoted by $k = 0$.

The following method is used to set the initial values of state variables of the controller. The cost function to be minimized is the integral of the squares of state variables such as velocity and position and is given by

$$J = \sum_{k=0}^{\infty} x[k]^T Q x[k], \tag{3.47}$$

where $x[k] = [x_p[k]^T, x_c[k]^T]^T$ is the state variable and Q is the weighting coefficient matrix.

The cost function can be transformed to the square of the initial values by using a *discrete Lyapunov equation* given by

$$J = x[0]^T P x[0], \tag{3.48}$$

where $P = \begin{bmatrix} p_{11} & p_{12} \\ p_{12}^T & p_{22} \end{bmatrix}$ is a positive definite matrix which can be obtained from $A^T P A - P = -Q$. Since $J > 0$ and J is a second-order polynomial in $x_c[0]$, there exists a minimum value of J for $x_c[0]$. By differentiating J with respect to $x_c[0]$, we get

$$\frac{\partial J}{\partial x_c[0]} = 2p_{11}x_c[0] + 2p_{12}x_p[0] = 0. \tag{3.49}$$

As such,

$$x_c[0] = -p_{11}^{-1} p_{12} x_p[0]. \tag{3.50}$$

During mode switching, the state variables of the controller are set as $x_c[0]$ in (3.50). This implies that the transient characteristics of track-settling control can be improved by minimizing J.

The z-transform of (3.43) to (3.46) with non-zero initial conditions is

$$y[z] = \frac{N_r[z]}{D[z]}r[z] + \frac{N_p[z]}{D[z]}x_p[0] + \frac{N_c[z]}{D[z]}x_c[0], \tag{3.51}$$

where $D[z]$ and $N_r[z]$ are scalar polynomials in z. $D[z] = \det[zI - A]$ and $A = \begin{bmatrix} A_p - B_p D_c C_p & B_p C_c \\ -B_c C_p & A_c \end{bmatrix}$. $N_p[z]$ and $N_c[z]$ are $1 \times m$ and $1 \times n$ polynomials, respectively, and can be concatenated as

$$[N_p[z] \quad N_c[z]] = [C_p \quad 0]\text{adj}(zI - A)z. \tag{3.52}$$

The initial values can now be obtained by introducing an $n \times m$ real matrix K and the following relation given by

$$x_c[0] = K x_p[0]. \tag{3.53}$$

Substituting (3.53) into (3.51), we have

$$y[z] = \frac{N_p[z] + N_c[z]K}{D[z]}x_p[0]. \tag{3.54}$$

(3.54) shows that the characteristics of the z-transform of the sequence $y[k]$ with non-zero initial conditions can be shifted to the desired values by selecting appropriate values of K. This suggests that the transient characteristics after mode switching can be improved by *shifting the zeros* to cancel the undesirable poles in (3.54).

As such, the zeros are carefully selected to cancel the undesirable closed-loop poles, which slows down the transient responses. The closed-loop poles can be represented by λ_k with $k = 1, 2, \ldots, j$, and j is the order of the controller. IVC designers can decide which of the poles λ_k should be cancelled.

Substituting $z = \lambda_k$ into the numerator polynomial in (3.54) for each element of $x_p[0]$, j simultaneous equations for the unknown $K(n, k)$ matrices can be obtained. Each $K(n, k)$ can now be solved under the condition for either $j = n$ or resetting of the residual state variables of the controller when $n > j$. Examples of the transient responses obtained with or without using the IVC method are shown in Figures 3.15(a) and 3.15(b), respectively. From these figures, it can be seen that the IVC method reduces the overshoot and eliminates the sluggish response from the slow eigenvalues.

In this method, the undesirable poles can be cancelled, and uncancelled poles dominate the transient response. The remaining poles are not always the most desirable poles for the transient response, as the precise positioning mode should be designed not only considering the transient characteristics but also the steady-state behaviour. To meet this requirement, another IVC method is proposed by including an additional input during mode switching so that the desirable pole in the transient response can be included. For this method, the additional input r' is added *after* the controller output and is represented by

$$r'[z] = \frac{n[z]}{d[z]} x_p[0], \tag{3.55}$$

with $x_c[0] = 0$. $n[z]$ and $d[z]$ are polynomials in z and $d[z]$ is stable. The transfer function between $r'[z]$ and $y[z]$ is assumed to be $\frac{Nr[z]}{D[z]}$, where $Nr[z]$ and $D[z]$ are polynomials in z. (3.51) can then be written as

$$y[z] = \frac{N_r[z]n[z] + N_p[z]d[z]}{D[z]d[z]} x_p[0]. \tag{3.56}$$

The transfer function of the right-hand side of (3.56) should be stable and proper and contains only the desirable poles ζ_i where $i = 1, 2, \ldots, l$. Since the roots of $D[z]$ do not always contain the desirable roots for the transient response, all the roots of $D[z]$ should be cancelled by zeros so that $d[z]$ contains only the desirable poles. There exists a condition to satisfy the above-mentioned pole-zero cancellation and interested readers are referred to [32] for more details.

The transient response for each eigenvalue is shown in Figure 3.16. The experimental and simulation results of the transient responses using the feed-forward type IVC with these eigenvalues are shown in Figures 3.17(a)–3.18(b). It should be noted that η in all figures are found from $y(s) = \frac{1}{s+a} x(s)$ and given by

$$\eta = e^{-aT}, \tag{3.57}$$

where T is the sampling period.

68

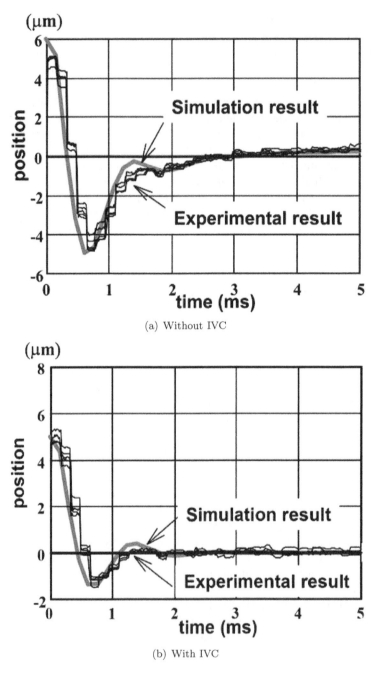

(a) Without IVC

(b) With IVC

FIGURE 3.15: Transient response by pole-zero cancellation.

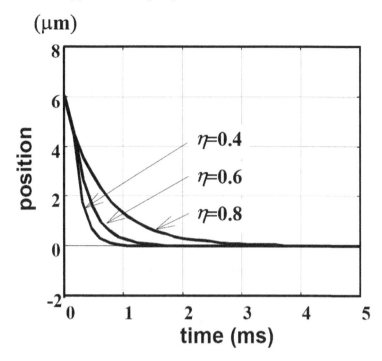

FIGURE 3.16: Simulated ideal impulse response of a first-order system.

The transient responses can be modified freely by assigning appropriate eigenvalues. In the case of $\eta = 0.2$, vibrations can be seen and are caused by excitation of mechanical resonant modes, since high frequency components in the control input are increased when shortening the transient response.

As such, it is important to take into account the accuracy of the model, calculation time, etc., when finding appropriate poles to be added to the IVC design. It should be noted that although the proposed IVC method modifies the set of poles from the transfer function of the initial state variables to the control variable for a desirable transient response, the structure of the controller remains *unchanged* and hence basic closed-loop characteristics such as stability are not compromised.

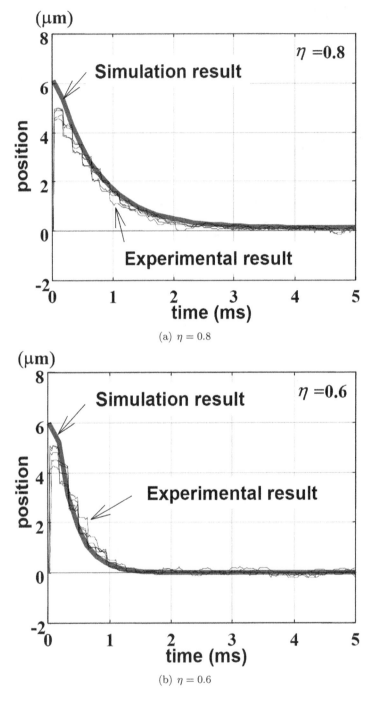

(a) $\eta = 0.8$

(b) $\eta = 0.6$

FIGURE 3.17: Transient response of IVC with feedforward input.

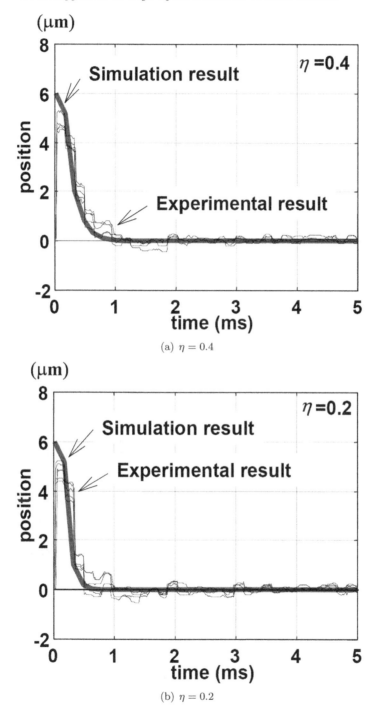

(a) $\eta = 0.4$

(b) $\eta = 0.2$

FIGURE 3.18: Transient response of IVC with feedforward input.

3.3.3 Optimal Design of Mode Switching Condition

The transient response after mode switching is influenced by the set of initial state variables of the plant such as position, velocity, and acceleration. As such, it is necessary to optimize the set of initial state variables during mode switching. As the number of initial state variables of the closed-loop system can be reduced to the number of state variables of the plant after the IVC design, it is possible to find a set of initial state variables of the plant which minimizes a certain performance index. In this section, the $\mathcal{H}_2$ norm is used as the index with

$$J_u \equiv \|Gx_p[0]\|_2^2, \tag{3.58}$$

where G is a transfer function from $x_p[0]$ to the control input. In this case, the set of initial state variables is designed to minimize the control input, and $x_p[0]$ which minimizes (3.58) can be calculated. The computational results for the $\mathcal{H}_2$ norm of velocity and acceleration are shown in Figure 3.19. Interested readers are kindly referred to [38] for more details.

In this case, the position during mode switching is fixed. In Figure 3.19, the bold line represents the set of initial state variables which optimizes the transient response of control input. The experimental results of transient responses of head position and current at (A), (B), and (C) in Figure 3.19 are shown in Figures 3.20(a)–3.21(c).

In these figures, many sets of experiments are carried out and the results are drawn in the same figures. The amplitude of current is minimum and the responses are fastest at (A), which corresponds to the mode switching with minimum $\mathcal{H}_2$ norm. It is worth noting that all the free parameters in (3.54) and (3.56) can be fixed using the above-mentioned IVC design, and thus it is verified that the proposed IVC design methodology can improve transient responses without affecting closed-loop characteristics such as stability.

There are many examples in various industries which utilize mode switching in motion control. This is obvious since various motions such as moving, positioning, actuating, picking, and placing actions, etc., are often required to be performed in a *single* sequence. In a chemical plant, for example, the operating mode is switched from an initial operation to its nominal operation and this is also a form of MSC. However, one issue of the IVC is the requirement of accuracy in measurement or estimation of the state variables of the plant during mode switching. In HDD head-positioning control systems, the IVC and optimal mode switching design have been implemented in actual products for many years, and their effectiveness is extensively verified. This means that the required accuracy of the measurement or estimation of the state variables can be handled.

The experimental results of position, velocity, and current waveforms during track-seeking, track-settling, and track-following are shown in Figure 3.22. In this figure, the waveforms of the bi-directional movement between two tracks are overwritten. It can be observed that the current waveform is saturated during acceleration, and the effects of back Electro-Motive Force (EMF) can also be seen. A larger current is available due to back EMF during deceleration. An exponential-like deceleration begins at a certain distance from the target position, and the head approaches the target position by following a smooth trajectory. Also, no overshoot or offset can be seen in Figure 3.22.

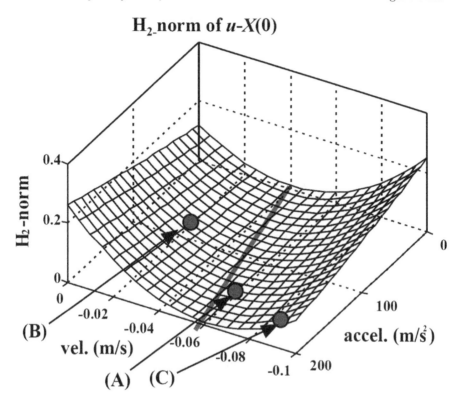

FIGURE 3.19: Optimal mode switching condition.

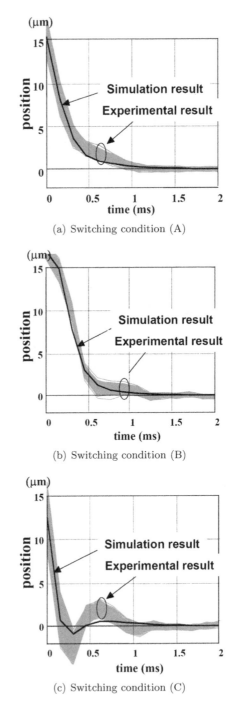

(a) Switching condition (A)

(b) Switching condition (B)

(c) Switching condition (C)

FIGURE 3.20: Transient response of head position.

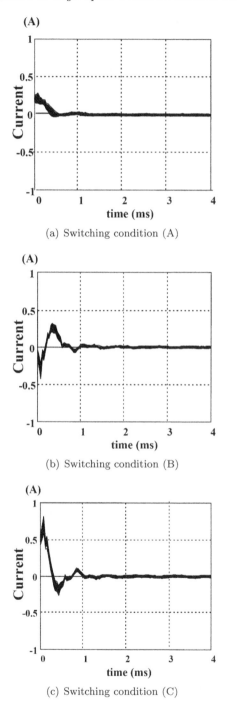

(a) Switching condition (A)

(b) Switching condition (B)

(c) Switching condition (C)

FIGURE 3.21: Transient response of current (experimental result).

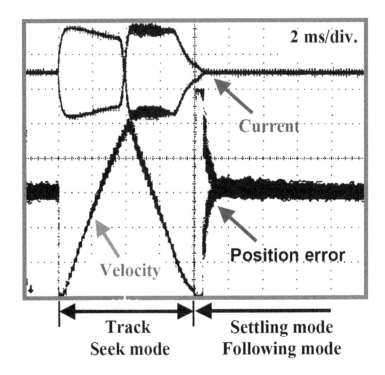

FIGURE 3.22: Time-domain waveform of head movement in HDDs.

3.4 Track-Following: Single- and Multi-Rate Control

The main objective during the track-following control mode is high-precision control of the magnetic head under the existence of disturbances and noise. In this section, the concepts of *single-rate* and *multi-rate* control systems are described.

3.4.1 Single-Rate Control

In this section, the main components of a single-rate track-following control system are detailed.

3.4.1.1 Introduction

In this section, examples of track-following control based on classical and modern control theories are explained. The design of the *lead compensator*,

Proportional-Integral (PI) controller, and *notch filter* for compensation of high frequency resonant modes based on classical control theory will be introduced.

3.4.1.2 Lead Compensator and PI Controller

The basic characteristics of the plant ignoring mechanical resonant modes can be given by

$$G(s) = \frac{K_g}{s^2}. \tag{3.59}$$

This is known as the *double-integrator* model, and is an unstable system as it has two poles at the origin. In order to stabilize this system with the Type-I servo characteristics, the plant requires *lead compensation* and *integral compensation* according to classical control theory. In this part, examples of the use of the lead compensator and PI controller to meet the above requirements will be provided.

The block diagram of the control system is shown in Figure 3.23. The

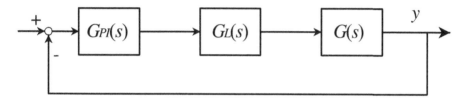

FIGURE 3.23: Block diagram of a control system.

transfer functions of the lead compensator and the PI controller are

$$G_L(s) = \frac{\alpha s + 2\pi F_L}{s + \alpha 2\pi F_L} \tag{3.60}$$

and

$$G_{PI}(s) = 1 + \frac{2\pi F_i}{s}, \tag{3.61}$$

where F_L and α are the center frequency and the upper to lower frequency ratio of the lead compensator, respectively. F_i is the cut-off frequency of the PI controller. The Bode plot of the compensator is shown in Figure 3.24.

The objective of track-following control in HDDs is to suppress the positioning error to within a specified value when subjected to steady-state disturbances. As described in the previous chapter, the disturbances are larger at the lower frequency region. As such, the control system is required to have a large bandwidth in order to improve the control performance. In HDD control, the bandwidth is defined by the frequency with which the gain of the open loop transfer function crosses the 0 dB line. This frequency is also known as the *gain crossover frequency.*

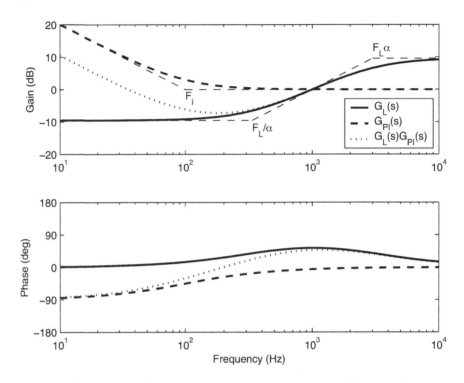

FIGURE 3.24: Bode plot of lead compensator and PI controller.

In general, there are many trade-offs between the control bandwidth and the robustness of the system. In classical control theory, robustness is usually measured by the *phase margin* and *gain margin*. For HDDs, the latter is more important than the former as the phase perturbation is not as large compared to the gain perturbation within the control bandwidth. The criteria of the phase margin and gain margin are decided not only by the stability of the system, but also by the *sensitivity transfer function*. The phase and gain margins represent the distance from the $(-1, 0)$ on the *Nyquist plot*, which is also the inverse of the gain of the sensitivity transfer function. As such, the gain of the sensitivity transfer function is a direct performance index for the head-positioning control system in HDDs, and the main objective of track-following control is to minimize the positioning error under the stationary disturbances. Motivated by these concepts, we target to optimize the control system from this point of view.

It rarely occurs that the track-following control performance of HDDs is restricted by the control input, thereby requiring advanced control methods such as the Linear Quadratic Regulator (LQR), which minimizes the square of sum of the control input and outputs as the performance index. In order to improve the control bandwidth, it is also important to consider the phenomenon known as the *waterbed effect*.

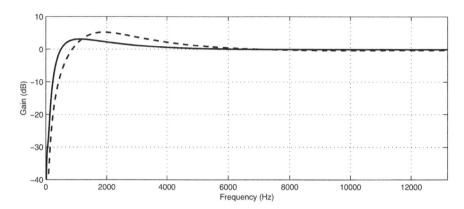

FIGURE 3.25: Sensitivity functions with two different control bandwidths illustrating the waterbed effect.

Consider an SISO Linear Time-Invariant (LTI) system with a stable open loop transfer function $L(s)$. The sensitivity transfer function is given by $S(s) = \frac{1}{1+L(s)}$. The closed-loop transfer function is stable, the following equation known as *Bode's Integral Theorem* holds

$$\int_0^\infty \ln |S(j\omega)| d\omega = -\frac{1}{2}\pi K_s, \qquad (3.62)$$

where $K_s = \lim_{s\to\infty} sL(s)$. If the relative order of $L(s)$ is more than or equal to two, $K_s = 0$. This means that when the gain of the sensitivity transfer function is plotted on the linear x-scale and logarithmic y-scale, as shown in Figure 3.25, the areas above and below the 0 dB line are equal. In other words, if the disturbances are *suppressed in a certain frequency range*, the disturbances will be *amplified in another frequency range*. The Position Error Signal (PES) spectrum corresponds to the product of the disturbance spectrum and the gain of the sensitivity transfer function. As such, it is necessary to decide the frequency ranges where the disturbances should be attenuated or be allowed to be amplified. A rule-of-thumb is to lower the gain of the sensitivity transfer function at the frequency range where the disturbances are large and vice versa. However, flutter disturbances often exist around the frequency where the "hump" of the sensitivity transfer function (corresponding to a frequency range where the gain of the sensitivity transfer function is above unity) is in the disturbance spectrum.

In almost all cases today, controllers are realized and implemented as *digital controllers*. In this book, we consider only digital controllers and discrete-time systems, the basic formulae for discretization are provided. Interested readers are referred to [40] for more details.

The continuous-time state-space equations of an LTI system are given by

$$\dot{x}(t) = Ax(t) + Bu(t), \tag{3.63}$$
$$y(t) = Cx(t), \tag{3.64}$$

where x is the state vector, y is the output vector, and u is the input vector. A, B, and C are the state matrices. Using a Zero-Order Hold (ZOH) and a sampler, the discrete-time state-space equations of the system are

$$x[k+1] = A_d x[k] + B_d u[k], \tag{3.65}$$
$$y[k] = C_d x[k], \tag{3.66}$$

where

$$A_d = e^{AT_s}, \tag{3.67}$$

$$B_d = \int_0^{T_s} e^{A\tau} d\tau\, B, \tag{3.68}$$

$$C_d = C, \tag{3.69}$$

and T_s is the sampling period.

Applying z-transform to the discrete-time state-space equations, we have

$$zx[z] = A_d x[z] + B_d u[z] \tag{3.70}$$
$$y[z] = C_d x[z]. \tag{3.71}$$

The function relating the z-transform of the input signal to the z-transform of the output signal is given by

$$y[z] = C_d[zI - A_d]^{-1} B_d u[z], \tag{3.72}$$

which is also known as the *pulse transfer function*.

The discrete-time controller can also be obtained by applying a transformation to a controller designed in the continuous-time domain. This method is known as *digital realization*. The *bilinear transformation* using trapezoidal approximation of the time integral for one sample period is commonly used for digital realization, especially for controllers like the lead compensator. This transformation can be represented by

$$s \to \frac{2}{T_s} \frac{1 - z^{-1}}{1 + z^{-1}}. \tag{3.73}$$

On the other hand, the discrete-time PI controller is realized with a digital integrator and is given by

$$G_{PI}[z] = 1 + \frac{\omega_i}{1 - z^{-1}} \tag{3.74}$$

or

$$G_{PI}[z] = 1 + \frac{\omega_i z^{-1}}{1 - z^{-1}}. \tag{3.75}$$

These equations correspond to the *backward approximation* (or *Euler approximation*) and a *forward approximation*, respectively. Their approximations in discrete-time *difference equations* are given by

$$y[k] = y[k - 1] + u[k] \tag{3.76}$$

or

$$y[k] = y[k - 1] + u[k - 1], \tag{3.77}$$

respectively.

There is no significant difference between the PI controllers, as can be seen from the above equations. The difference is just in the update of timing, and the designer can use whichever form is convenient during implementation. Next, we proceed to design a controller using the lead compensator and an integrator for the HDD Benchmark Problem. The control performance for various parameters will also be shown in simulation.

In the classical control theory framework, the main specifications used for controller design are the control bandwidth, phase and gain margins, peak and suppression gains at the low frequency of the sensitivity transfer function, etc. These translate to design parameters in terms of open-loop gain crossover frequency, center frequency, upper to lower frequency ratio of the lead compensator, and integral time of the integrator. The effects of the parameters in both time and frequency responses are shown in Figures 3.26 and 3.27.

The upper to lower frequency ratio of the lead compensator is a parameter which determines the amount of phase lead. In continuous-time systems, the phase lead will be maximum at the center frequency if there is no time delay, and hence the center frequency of the lead compensator should coincide with the gain crossover frequency. However, for discrete-time systems, the center frequency is usually chosen to be greater than the gain crossover frequency to increase the upper to lower frequency ratio, as the phase delay caused by sampling and calculation is significant. As mentioned previously, the phase margin is less important than the gain margin when determining closed-loop stability. As a matter of fact, the phase margin is regarded as a parameter for determining the peak gain of the sensitivity transfer function. A large phase margin will result in a smaller peak gain in the sensitivity transfer function, which in turn implies reduced disturbance attenuation in the low frequency region. If the disturbance spectrum is known, the PES spectrum can be estimated by using the sensitivity transfer function and the controller parameters can be easily tuned based on the estimated PES. An example of PES spectrum and the corresponding 6σ value of Non-Repeatable Run-Out (NRRO) are shown in Figure 3.28. The effects of the different controllers to positioning accuracy can be clearly seen.

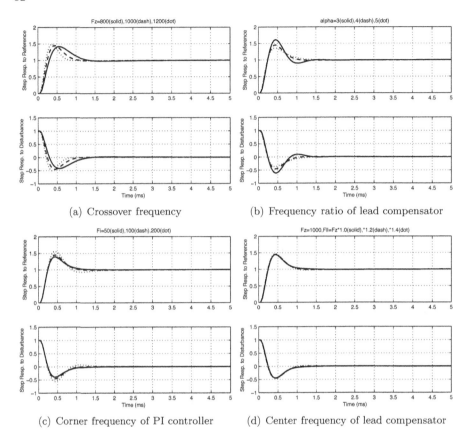

(a) Crossover frequency

(b) Frequency ratio of lead compensator

(c) Corner frequency of PI controller

(d) Center frequency of lead compensator

FIGURE 3.26: Time responses of step reference and disturbance.

3.4.1.3 Notch Filter

In general, the actuators in HDDs have many resonant modes in the high frequency range with large parametric variations. As such, it is necessary to compensate for the gain or phase of the plant to avoid instability of the closed-loop system. A Low Pass Filter (LFP) can be used for gain compensation, but *multi-stage notch filters* are preferred as they minimize the phase delay, resulting in better performance of the control systems.

In this section, we show an example where the most commonly used second-order notch filters are connected in series. The continuous-time transfer function of a typical second-order notch filter is given by

$$G_N(s) = \frac{s^2 + d2\zeta\omega_N s + \omega_N^2}{s^2 + 2\zeta\omega_N s + \omega_N^2}, \qquad (3.78)$$

where ω_N is the center frequency. ζ and d are related to width and depth of

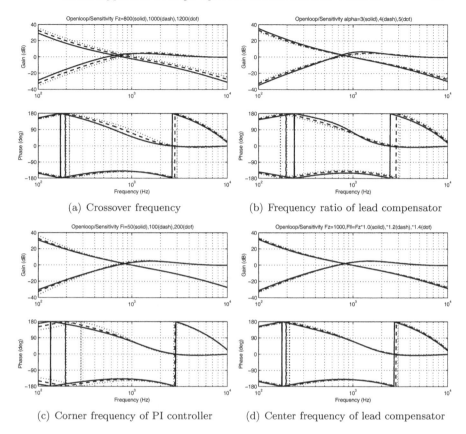

(a) Crossover frequency

(b) Frequency ratio of lead compensator

(c) Corner frequency of PI controller

(d) Center frequency of lead compensator

FIGURE 3.27: Frequency responses of open-loop and sensitivity transfer functions.

the notch, respectively. The frequency responses of $G_N(s)$ are shown by the dashed lines in Figure 3.29.

To implement it as a digital filter, the *bilinear transform* is used as a discrete-time transformation. In the case of notch filters, the shifts in radial frequency affect the characteristics of the filters significantly. As such, ω_N should be converted using *frequency prewarping* accordingly to

$$\omega'_N = \frac{2}{T_s} \tan\left(\omega_N \frac{T_s}{2}\right). \tag{3.79}$$

The result of the discretization of $G_N(s)$ is

$$G_N(z) = \frac{(1 - d\zeta k + k^2)z^{-2} + (2k^2 - 2)z^{-1} + (1 + d\zeta k + k^2)}{(1 - \zeta k + k^2)z^{-2} + (2k^2 - 2)z^{-1} + (1 + \zeta k + k^2)}, \tag{3.80}$$

where $k = \tan\left(\omega_N \frac{T_s}{2}\right)$.

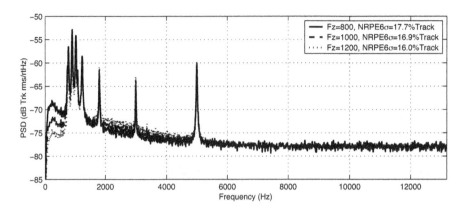

FIGURE 3.28: PES spectra.

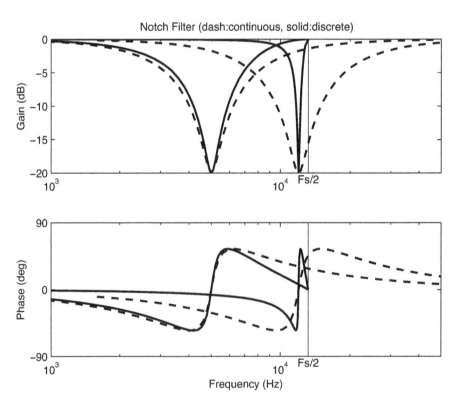

FIGURE 3.29: Notch filter and effects of discretization using bilinear transformation.

If the center frequency is close to the Nyquist frequency, the width of the digital notch filter will be narrow, as can be seen in Figure 3.29. This width can be adjusted by varying ζ. By connecting multiple second-order filters in series, a filter which lowers the peak gain of the resonant characteristics can be designed, as shown in Figure 3.30. When tuning the notch filters, the Bode and Nyquist plots can be combined to decide which parameters provide sufficient stability margins by taking into account the perturbations of the plant directly. Adding a notch filter increases the phase delay and reduces the achievable control bandwidth. In order to minimize the phase delay, a large number of stages of notch filters is required to bound the resonant characteristics of the plant with perturbation as tightly as possible. The Bode and Nyquist plots of a control system with plant perturbation are shown in Figure 3.31, and the closed-loop systems are stable in all cases. It is worth noting that the `c2d.m` command in MATLAB® with the `'prewarp'` option is used in this example, but the `'matched'` option using pole-zero matching method is better for higher order systems.

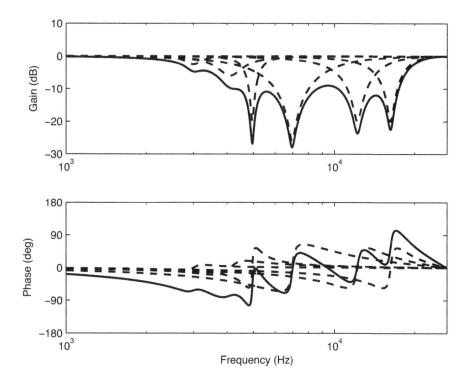

FIGURE 3.30: Bode plots of multi-stage notch filters.

3.4.1.4 Observer State Feedback Control

In this section, an observer state feedback control based on the modern control theory is explained. According to modern control theory, the closed-loop

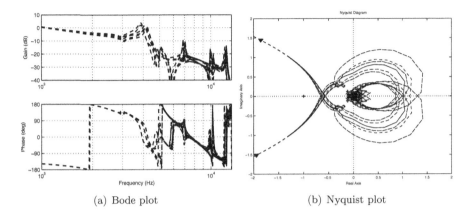

(a) Bode plot   (b) Nyquist plot

FIGURE 3.31: Frequency responses of the perturbed open-loop model.

poles can be arbitrary placed by a feedback gain if the plant is *controllable*. However, it is not always possible to measure all the states of the system. But if the system is *observable*, all the states can be observed and used for feedback control using a *state observer*. This is known as *observer state feedback control*, and the overall block diagram is shown in Figure 3.32.

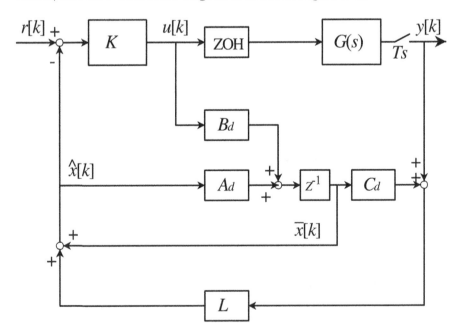

FIGURE 3.32: Block diagram of observer-based state feedback control.

The plant model for the state observer is usually a simplified model. In this case, the HDD actuator is approximated by a double integrator with time delay. A bias force is added to give the Type-I servo characteristics and the state equations are given by

$$
\begin{aligned}
\dot{x}(t) &= Ax(t) + Bu(t - T_d), & (3.81)\\
y(t) &= Cx(t) + Du(t - T_d), & (3.82)
\end{aligned}
$$

where

$$
x = \begin{bmatrix} x_1 & x_2 & x_3 \end{bmatrix}^T, \tag{3.83}
$$

and

$$
A = \begin{bmatrix} 0 & 1 & 0 \\ 0 & 0 & K_g \\ 0 & 0 & 0 \end{bmatrix}, B = \begin{bmatrix} 0 \\ K_g \\ 0 \end{bmatrix}, C = \begin{bmatrix} 1 & 0 & 0 \end{bmatrix}, D = \begin{bmatrix} 0 \end{bmatrix}. \tag{3.84}
$$

The system is then discretized at a sampling time T_s where $T_d = mT_s$ with $0 \le m < 1$. For simplicity, the units of velocity and acceleration are track$/T_s$ and track$/T_s^2$, respectively, and the units of both control input and bias force are assumed to be same as the acceleration.

The state-space equations of the discretized system can be given by

$$
\begin{aligned}
x[k + 1] &= A_d x[k] + B_d u[k], & (3.85)\\
y[k] &= C_d x[k], & (3.86)
\end{aligned}
$$

where $x = \begin{bmatrix} x_1 & x_2 & x_3 & x_4 \end{bmatrix}^T$,

$$
A_d = \begin{bmatrix} 1 & 1 & \frac{K_g}{2} & B_{11} \\ 0 & 1 & K_g & B_{21} \\ 0 & 0 & 1 & 0 \\ 0 & 0 & 0 & 0 \end{bmatrix}, B_d = \begin{bmatrix} B_{12} \\ B_{22} \\ 0 \\ 1 \end{bmatrix}, \tag{3.87}
$$

$$
C_d = \begin{bmatrix} 1 & 0 & 0 & 0 \end{bmatrix}, \tag{3.88}
$$

$$
\begin{bmatrix} B_{11} \\ B_{21} \end{bmatrix} = \begin{bmatrix} 1 & 1 - m \\ 0 & 1 \end{bmatrix} \begin{bmatrix} \frac{m^2 K_g}{2} \\ m K_g \end{bmatrix} = \begin{bmatrix} \frac{(2m - m^2) K_g}{2} \\ m K_g \end{bmatrix}, \tag{3.89}
$$

$$
\begin{bmatrix} B_{12} \\ B_{22} \end{bmatrix} = \begin{bmatrix} \frac{(1 - m)^2 K_g}{2} \\ (1 - m) K_g \end{bmatrix}, \tag{3.90}
$$

and

$$
\begin{bmatrix} B_{11} + B_{12} \\ B_{21} + B_{22} \end{bmatrix} = \begin{bmatrix} \frac{K_g}{2} \\ K_g \end{bmatrix}. \tag{3.91}
$$

The state variables x_1, x_2, and x_3 are position, velocity, and bias force, respectively, and x_4 is the control input at the previous sample. An additional state is included when the time delay is considered and discretized. The equivalent time delay not only includes the delay from the time when the position is detected to the time the control input is updated, but also includes the phase delay of the current amplifier of the VCM and that from the notch filters. In other words, this means that overall effects can be approximated by a double integrator and time delay. The Bode plot of the plant model is shown in Figure 3.33.

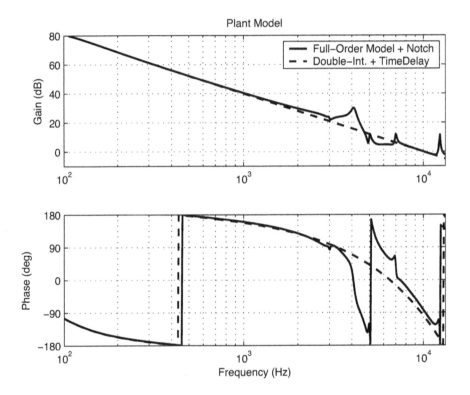

FIGURE 3.33: Frequency responses of full-order and reduced-order plant models.

There are several ways to realize a discrete-time state observer. In this section, we describe a *full-order current state observer* which is given by [40]

$$\hat{x}[k] = \bar{x}[k] + L(y[k] - C_d\bar{x}[k]), \qquad (3.92)$$
$$\bar{x}[k+1] = A_d\hat{x}[k] + B_du[k], \qquad (3.93)$$

where $\bar{x}$ represents the predicted values, $\hat{x}$ represents the estimations which are corrected at every sample time based on the difference between the detected value $y(k)$ and the predicted value, and L is the observer gain. The response of the observer is expressed in terms of the estimation error and is given by

$$\hat{e}[k+1] = (A - LCA)\hat{e}[k], \tag{3.94}$$

and the observer gain L should be designed to make the estimation error $\hat{e}[k] = \hat{x}[k] - x[k]$ converge to zero.

State feedback is then expressed as

$$u[k] = -K(\hat{x}[k] - r[k]), \tag{3.95}$$

where $r[k]$ is the reference state vector in which all the values should be zero during track-following.

Combining both the observer and state feedback controller, the overall system can be represented in state-space form as

$$\begin{aligned}
\bar{x}[k+1] &= [(A_d - B_dK)(I - LC_d)]\bar{x}[k] \\
&+ \begin{bmatrix} (A_d - B_dK)L & B_dK \end{bmatrix} \begin{bmatrix} y[k] \\ r[k] \end{bmatrix}, \tag{3.96}
\end{aligned}$$

$$u[k] = [-K(I - LC_d)]\bar{x}[k] + \begin{bmatrix} -KL & K \end{bmatrix} \begin{bmatrix} y[k] \\ r[k] \end{bmatrix}. \tag{3.97}$$

The time and frequency responses to a disturbance or reference can now be obtained.

3.4.1.5 Pole Placement Technique

There are various ways of designing the observer and state feedback gain. In this section, the *pole placement* technique will be described.

Pole placement is a method which places the closed-loop poles of the system at specific locations on the z-plane in order to achieve certain desired responses. In this case, the frequencies and damping factors are used as guidelines for obtaining the desirable responses. It is a well-known fact that the stable left-half side of the s-plane is projected into a unit circle on the z-plane, and the frequencies and damping factors for the complex poles at various locations within the unit circle are shown in Figure 3.34. As such, it is easy to place the poles using the frequencies and damping factors. The location of the pole on the z-plane expressed in terms of the pole frequency and damping factor is given by $E = \exp(-\omega T_s)$ for the case of the real pole and $E = \exp\left\{\left(-\zeta\omega \pm j\omega\sqrt{1-\zeta^2}\right)T_s\right\}$ for the case of the complex pole.

When the plant is controllable and observable, the poles of state feedback and state observer can be placed arbitrarily. The theoretical solutions can be obtained using Ackerman's formula [40], or by using the `place.m` command in MATLAB® though there are limitations to this command and it can provide inaccurate answers at times.

The pole placement is basically a design method based on the time response of the observer and feedback poles, while classical control theory based

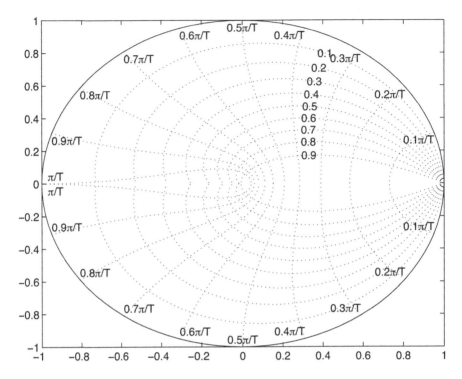

FIGURE 3.34: Pole-zero map and damping ratio on the z-plane.

design shapes the frequency responses directly. However, it is important to consider the response to steady-state disturbances in the frequency domain for track-following control in HDDs. As such, it is necessary to take note of the frequency response even when the controller is designed using the pole placement method.

The design parameters are the frequency and damping factor of the complex poles of the position and velocity observer, pole frequency of the bias observer, and the frequency and damping factor of the poles of state feedback. The pole for the time delay term should be set to zero. Moreover, the feedback gain for the bias term should also be zero, and the basic control performance will depend on the state feedback poles. In other words, the frequencies of the state feedback poles will determine the control bandwidth, and the damping factor will determine the peak gain and low frequency suppression capabilities of the sensitivity transfer function. The complex poles of the observer are considered as parameters to adjust the trade-offs between the high frequency gain and phase delay.

In fact, the observer state feedback controller for the double integrator corresponds to a standard *Proportional-Integral-Derivative (PID) controller* with an additional second-order LPF. The frequencies and damping factors of

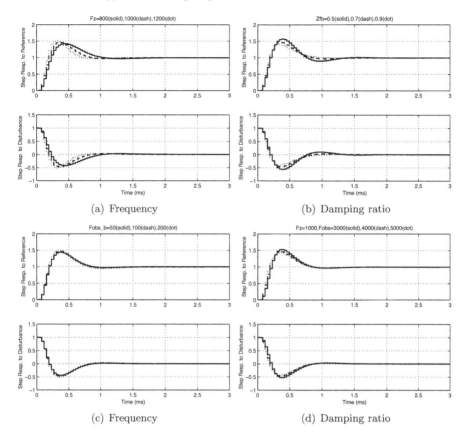

(a) Frequency   (b) Damping ratio

(c) Frequency   (d) Damping ratio

FIGURE 3.35: Time responses of step reference and disturbance using state feedback (upper) and estimator (lower).

the complex poles of the position and velocity observer correspond to those of a second-order LPF, and the pole frequency of the bias observer corresponds to the frequency of the integrator of the PI controller. As such, the method is essentially the *same* when compared to classical control theory design. However, this is only true in terms of the disturbance response. For the reference response, the use of the observer has the advantage of having an internal model to improve the performance. In addition, since information on position, velocity, and bias states is always available, it is convenient to set the initial values of the states to improve the transient response.

The responses and positioning errors of different examples of controller design using various design parameters for the HDD Benchmark Problem are shown in Figures 3.35–3.37.

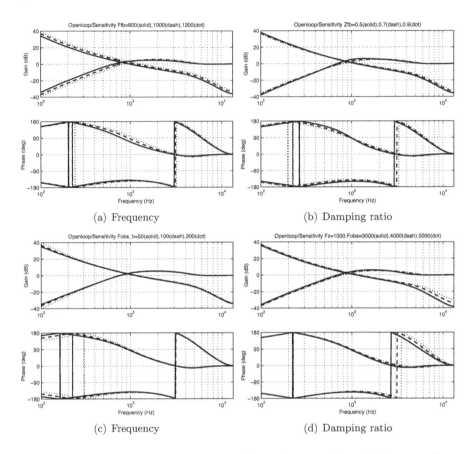

(a) Frequency

(b) Damping ratio

(c) Frequency

(d) Damping ratio

FIGURE 3.36: Frequency responses of open-loop and sensitivity transfer functions using state feedback (upper) and estimator (lower).

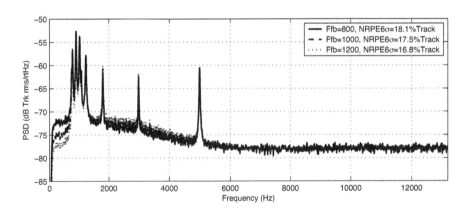

FIGURE 3.37: PES spectra.

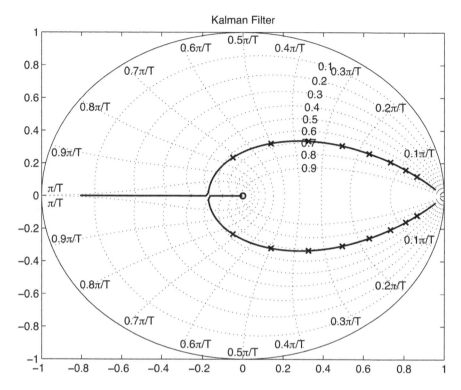

FIGURE 3.38: Root locus using the Kalman filter design.

3.4.1.6 Optimal Control Design

In modern control theory, the *Linear Quadratic Gaussian* (LQG) is an optimal control design method which combines the *Kalman filter* and the *Linear Quadratic Regulator* (LQR).

The root locus of the design using the Kalman filter is shown in Figure 3.38. Under certain assumptions, it can be seen that the damping factor is always around 0.7 as the frequency changes, which is similar to the case of the LQR, as shown in Figure 3.39. In general, the LQG provides sufficiently good but not necessarily optimal solutions. The optimal design theory can also be applied to the design of parameters for state feedback to reduce the dimension of the design parameter space. The parameters for state feedback control can first be obtained using an optimal control design such as the LQG, after which the pole placement method can be used to adjust the parameters.

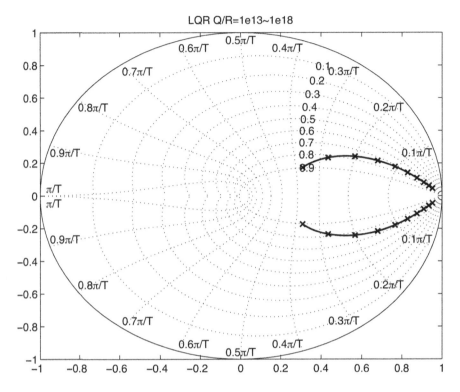

FIGURE 3.39: Root locus using the LQR design.

3.4.2 Multi-Rate Control

The PES has a fixed sampling rate in HDDs. As such, *multi-rate control* is commonly employed to overcome the limitation of a slow or fixed sampling rate of the controlled variable. In this section, the fundamentals of multi-rate control are presented.

3.4.2.1 Introduction

The position of the head is obtained by the PES which is embedded on the disk surface in HDDs. The sampling period is determined by the number of PES data and the rotational speed of the disk. As such, the sampling period cannot be chosen to be arbitrarily small to improve control performance. However, the control period does not have such a restriction, and it can be chosen to be very small as long as it is within the hardware limitation. With the recent advances in microprocessor and DSP technologies, it is easy to achieve control periods of half or a quarter of the sampling period of PES.

Currently, digital control is commonly used in HDDs. In high-end HDDs, a higher sampling frequency is used to maximize the control performance even though this decreases the format efficiency slightly. On the other hand, it is difficult for low-end and/or mobile HDDs to use such a high sampling frequency due to the need to maximize data capacity and/or reduce power consumption.

If the mechanical resonant modes of the plant exist above the Nyquist frequency, it is difficult to design a digital controller which takes into account these resonant modes since they may be excited by the control input. In this case, a notch filter is commonly used to cope with this problem so that the frequency components of the control input around the resonant frequencies are reduced. Recently, the notch filter was implemented as a digital filter whose sampling rate is N times *faster* than that of PES. A control system that has multiple sampling periods is referred to as a *multi-rate control system*. In HDDs, the multi-rate control technique plays an important role in maximizing the control performance under the constraint of the sampling frequency of PES. Conversely, a digital control system that has a single sampling period only is referred to as a *single-rate control system*.

Other than HDDs, multi-rate control is also suitable for many industrial applications. For example, it is common that each sensor has a different sampling period during sensor fusion. However, all the samples are usually *resampled* in order to be implemented as a single-rate control system. In this case, the control performance will be improved with the application of multi-rate control. In fact, multi-rate control can achieve a control performance that cannot be achieved using continuous or single-rate control, but this is outside the scope of our discussion and is omitted here for brevity.

3.4.2.2 Problem Formulation

In general multi-rate control systems, the sampling periods differ between input and output signals and also among input and output channels. As the general formulation of multi-rate control systems is mathematically involved, it is assumed that the plant is a SISO system and the sampling period of the plant output is an integer multiple of the plant input for simplicity but without loss of generality.

We define a continuous-time n^{th}-order state-space equation as

$$\dot{x}(t) = A_c x(t) + B_c u(t), \quad y = C_c x(t). \tag{3.98}$$

To discretize the system represented in (3.98) with a sampling period, a ZOH is used and is given by

$$u(kT_u + \theta) = u[k], \quad \theta \in [0, T_u). \tag{3.99}$$

The discrete-time state-space equation is now given by

$$x_d[k+1] = A_d x_d[k] + B_d u[k], \tag{3.100}$$

where

$$A_d := e^{A_c T_u}, \quad B_d := \int_0^{T_u} e^{A_c \tau} B_c d\tau, \quad x_d[k] := x(kT_u). \tag{3.101}$$

It is assumed that the output of the system as depicted in (3.100) is measured once in M times, i.e., $T_y := MT_u$, where T_y is a sampling period of y. By introducing $k = Mi + j$ where $i = 0, 1, 2, \ldots$ and $j = 0, \ldots, M - 1$, the measurement of the output y can be represented as

$$y[i] := Cx_d[Mi].\tag{3.102}$$

The time chart of u and y is shown in Figure 3.40.

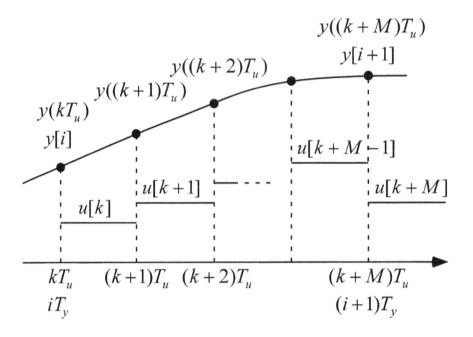

FIGURE 3.40: Input and output signals of the multi-rate system.

The multi-rate system as depicted in (3.100) is a *linear time-periodic system* because the output $y[i]$ is measured *only* when $j = 0$. It is difficult to handle this multi-rate system, and thus we describe the system in (3.100) at the slow sampling rate of T_y. Using simple algebra, an equivalent system can be obtained as

$$
\begin{aligned}
&x_d[M(i+1)] \\
&= A^M x_d[Mi] + \begin{bmatrix} A_d^{M-1}B_d, & A_d^{M-2}B_d, & \ldots, & B_d \end{bmatrix} \begin{bmatrix} u[Mi] \\ u[Mi+1] \\ \vdots \\ u[Mi+M-1] \end{bmatrix}.
\end{aligned}
\tag{3.103}
$$

Furthermore, by defining $x[i] := x_d[Mi]$ and

$$A := A_d^M, \quad B := [\ A_d^{M-1}B_d, \ A_d^{M-2}B_d, \dots, \ B_d\ ], \quad (3.104)$$

$$\underline{U}[i] := [\ u[Mi], \ u[Mi+1], \dots, \ u[Mi+M-1]\ ]^T, \quad (3.105)$$

we have

$$x[i+1] = Ax[i] + B\underline{U}[i], \quad (3.106)$$

$$y[i] = Cx[i]. \quad (3.107)$$

As such, SISO multi-rate systems with two sampling rates of T_u and T_y can be represented by a single-rate LTI system with M inputs and one output. The operation to convert $u[k]$ into $\underline{U}[i]$ is given by

$$\{u[0],\ u[1],\ u[2],\dots\} \mapsto \left\{ \begin{bmatrix} u[0] \\ \vdots \\ u[M-1] \end{bmatrix}, \begin{bmatrix} u[M] \\ \vdots \\ u[2M-1] \end{bmatrix}, \dots \right\},$$

$$(3.108)$$

and is called *discrete-time lifting* or *blocking*.

Since the lifted system as depicted in (3.106) is LTI, various control design methods can be applied to design the feedback controller $\underline{U} = \underline{K_d}y$. As (3.106) is a fictitious system, some discussion of the implementation issues of $\underline{K_d}$ is required.

$\underline{K_d}$ is a system with one input and M outputs. As such, the output of $\underline{K_d}$ includes signals from current time to $M-1$ steps ahead and depends on the current measurement $y[i]$ only. With this the *causality* condition is satisfied, which makes $\underline{K_d}$ implementable. On the other hand when $T_u = MT_y$, the feedback controller of the lifted system may not be realizable because the output of $\underline{K_d}$ at the current time may depend on future signals. In this case, $\underline{K_d}$ has to be obtained to satisfy some structural constraints for causality [41].

3.4.2.3 Multi-Rate Observer

In this section, a multi-rate control method is introduced based on a *multi-rate observer* which has been applied to actual HDDs. The multi-rate observer estimates the state variables of the plant at a higher sampling rate by using the control input at a fast sampling rate and the measurement output at a slow sampling rate. Using state feedback based on the estimated plant states, the control input is generated at the fast sampling rate. This is a simple and easy way to design a multi-rate controller.

It is assumed that the measured output can be obtained at the sampling period T_u of the control input, i.e., $y_d[k] := y(kT_u)$. The state observer is described by

$$\bar{x}_d[k] = A_d\hat{x}_d[k-1] + B_d u[k-1], \quad (3.109)$$

$$\hat{x}_d[k] = \bar{x}_d[k] + L(y_d[k] - C\bar{x}_d[k]), \quad (3.110)$$

where $\bar{x}_d$ and $\hat{x}_d$ are referred to as *predicted estimate* and *current estimate* of x_d, respectively. This type of observer is known as a *current estimator* [40], and provides a current estimate $\hat{x}_d[k]$ based on the current measurement $y_d[k]$. It is possible for $\hat{x}_d[k]$ to be more accurate than $\bar{x}_d$ as $\hat{x}_d[k]$ is updated by the most current measurement $y_d[k]$. The state feedback based on the current estimate cannot be implemented exactly because it is impossible to sample, perform calculations, and provide the output at the same time. However, it is possible to minimize computational delays by performing all calculations that do not directly depend on $y_d[k]$ in advance.

Since the plant is a multi-rate system, $y_d[k]$ can be measured only when $k = Mi$, i.e., when $j = 0$ is satisfied. As such, we can assume that $y_d[k]$ coincides with $C\bar{x}_d[k]$. This is the estimate of $y_d[k]$ when $k \neq Mi$ holds and we have

$$y_d[k] = C\bar{x}_d[k], \quad k \neq Mi. \tag{3.111}$$

Under this assumption, (3.109) turns to out be [42][43]

$$\bar{x}_d[k] = A_d\hat{x}_d[k-1] + B_d u[k-1], \tag{3.112}$$

where $k = Mi + j$ and

$$\hat{x}_d[k] = \begin{cases} \bar{x}_d[k] + L(y[i] - C\bar{x}_d[k]), & (k = Mi) \\ \bar{x}_d[k], & (k = Mi + j, \ j \neq 0) \end{cases} \tag{3.113}$$

By using discrete-time lifting, (3.112) can be described as an LTI system at a sampling period T_y as

$$\bar{x}[i] = A\hat{x}[i-1] + B\underline{U}[i-1], \tag{3.114}$$

$$\hat{x}[i] = \bar{x}[i] + L(y[i] - C\bar{x}[i]), \tag{3.115}$$

where $\bar{x}[i] := \bar{x}_d[Mi]$ and $\hat{x}[i] := \hat{x}_d[Mi]$. A, B, and $\bar{U}[i]$ have the same form as depicted in (3.104).

To look into the estimation error $e[i] := x[i] - \hat{x}[i]$, (3.114) is substituted into (3.115) to give

$$\hat{x}[i] = (I - LC)A\hat{x}[i-1] + (I - LC)B\underline{U}[i-1] + Ly[i]. \tag{3.116}$$

The lifted plant with the sampling period T_y can be described as

$$x[i] = Ax[i-1] + B\underline{U}[i-1]. \tag{3.117}$$

From (3.116) and (3.117), we have

$$e[i] = (A - LCA)e[i-1]. \tag{3.118}$$

It follows from (3.118) that the estimation error will be asymptotically zero by choosing L so that $(A - LCA)$ is Hurwitz if (A, CA) is observable. Finally, the multi-rate control is derived using state feedback based on the current estimate $\hat{x}_d[k]$ as

$$u[k] = -F\hat{x}_d[k]. \tag{3.119}$$

Next, we introduce a multi-rate observer that has been proposed in [44]. Since $y[i]$ is only obtained once in M updates of $u[k]$, the estimation is also updated once in M times. If the estimation error is large, $\hat{x}_d[k]$ and the control input $u[k]$ will change abruptly when $k = Mi$. This phenomenon may excite high frequency resonant modes of the plant. In particular when $T_u \ll T_y$, the control input excites the high frequency resonant modes above the Nyquist frequency of T_y, and the aliasing effect may degrade the control performance. As such, the following equation has been proposed in [44] in order to obtain a smooth estimation during the sampling interval of T_y with

$$\hat{x}_d[k] = \bar{x}_d[k] + L_j(y[i] - C\bar{x}[i]), \tag{3.120}$$

where $k = Mi+j$ and L_j is the observer gain for each j. It follows from (3.120) that the estimation error $y[i] - C\bar{x}[i]$ at $j = 0$ can be corrected not only when $j = 0$ but also when $j = 0, \ldots, M - 1$. Therefore, the abrupt change of the control input when $j = 0$ can be prevented.

$\hat{x}[i]$ is the behavior of $\hat{x}_d[k]$ at $k = Mi$ and is determined by the characteristics equation with

$$\det\left[zI - A_d^M + \sum_{h=1}^{M}\left(A_d^h L_{M-h}\right)C\right] = 0. \tag{3.121}$$

By assuming that L_j is described as $L_j = a_j\bar{L}$ where $a_j \in \mathcal{R}$ and

$$\bar{L} = \left[\sum_{h=1}^{M}\left(a_{M-h}A_d^h\right)\right]^{-1} A_d^M L, \tag{3.122}$$

we can show that (3.121) coincides with (3.118). As such, L_j can be determined by (3.122) using the L obtained by (3.118). Note that a_j has to be determined when $\bar{L}$ and L_j are calculated, and $a_j = 1$ for all j is proposed in [44].

The simulation and experimental results in [44] show that the multi-rate controller based on the multi-rate observer using (3.120) generates a smooth control input for both step reference and disturbance responses.

3.5 Episode: Development of IVC Design Method in Industry

I returned to a research center after finishing the development of my company's first 3.5″ digital control HDD in 1989. Then, my highest priority was to improve the transient response of this HDD during settling, which was *worse* than conventional HDDs utilizing analog circuits for both continuous servo control and PES. At that time, digital controllers using discrete-time based control were directly converted from continuous-time based control, and it was common that their performance was much inferior to the original. Although mode switching from track-seeking mode to track-settling mode has been a popular method since the appearance of analog controllers, I had much trouble writing program codes for the track-settling mode, especially for the *first* calculation of the track-settling mode after mode switching. At the first sample, the control output u is calculated by the output equation, and the state variables $x_c[k+1]$ are then calculated by the state equations which will be used during the next sample. However, in the case of mode switching, the question was: *What are the values of $x_c[k]$ for calculating the output equation given by*

$$
\begin{aligned}
x_c[k+1] &= A_c x_c[k] + B_c(r[k] - y[k]), \\
u[k] &= C_c x_c[k] + D_c(r[k] - y[k])?
\end{aligned}
$$

Traditional wisdom in control textbooks told me that the values of $x_c[k]$ should be zero since controller calculation always starts from the initial state. In fact, we set the values to zero during mode switching in HDDs at the beginning, even when the state was *not* a steady-state during mode switching. I was really annoyed by this issue. What should be the value of the input? However, I quickly changed my mindset from the negativity of "I do not know what should be the value of the input" to the *positivity* of "I can input any value!" This made me noticed that $x_c[k]$ during mode switching is a *free* parameter, and control engineers can now find appropriate performance indices to optimize $x_c[k]$. It was then that I confirmed that the transient response must be improved by using the values calculated from these indices.

Control theory is quite useful as the Lyapunov equation was proposed to solve this problem. The result was outstanding and the transient response was dramatically improved. From the implementation point of view, it was also quite easy since it reduced to just adding a simple calculation during the first sample after mode switching. When I suggested this set of initial values calculated by the IVC method to the HDD development engineer, his boss never believed in this set of values and asked him to search for *much better* numbers using the so-called "carpet-bombing" method. The result was actually quite straightforward and predictable; *all* experimental data other than those suggested by me was much worse. I hope his boss finally realized the importance of control theory.

The next step was to investigate how this idea could contribute to the field of control theory. Certainly, there were already several papers in the current literature which mentioned that the initial value should be set to the integrator of the controller at the beginning of the step response from steady-state. In HDD servo systems, a similar approach was already used for the analog circuit controller where a certain voltage is injected into the integrator during mode switching. However, this study was only presented as a one page proceedings in a domestic conference. I did a thorough literature survey of papers and patents pertaining to IVC and its related fields, and visited many university professors for discussions and suggestions. It was after that that I confirmed this idea could be built as a *new design methodology* for servo control. Development of the IVC method was completed in 1998 and since then, the IVC had been applied to not only HDDs but also other mechatronic products such as Galvano mirror control in laser drilling machines, etc.

Three lessons were learnt from this experience. First, good "seeds", ideas, and problems can be found at the product development site. Second, knowledge of control theory is certainly great and important. Last, it is important that even engineers in industries spend time forming their developed technologies as methodologies which contribute to development in control theory, in addition to daily product development activities in the company. This extra effort will definitely help them advance to the next level.

Takashi Yamaguchi

Bibliography

[1] N. Suda, *PID Control*, Asakura Publishing, 1996 (in Japanese).

[2] M. Araki and H. Taguchi, "Two-Degrees-of-Freedom PID Controllers," *International Journal of Control, Automation, and Systems*, Vol. 1, No. 4, pp. 401–411, 2003.

[3] M. Tomizuka, "Zero Phase Error Tracking Algorithm for Digital Control," *ASME Journal of Dynamic Systems, Measurement and Control*, Vol. 109, pp. 65–68, 1987.

[4] M. Tomizuka, "On the Design of Digital Tracking Controllers," *ASME Journal of Dynamic Systems, Measurement and Control*, Vol. 115, pp. 412–418, 1993.

[5] M. Kobayashi, T. Yamaguchi, I. Oshimi, Y. Soyama, and H. Hirai, "Multirate Zero Phase Error Feedforward Control for Magnetic Disk Drives," in *Proceedings of the Conference on Information, Intelligence and Precision Equipment (IIP'98)*, Vol. 98, No. 26, pp. 21–22, 1998 (in Japanese).

[6] B. Haack and M. Tomizuka, "The Effect of Adding Zeros to Feedforward Controllers," *ASME Journal of Dynamic Systems, Measurement and Control*, Vol. 113, pp. 6–10, 1991.

[7] D. Torfs, J. De Schutter, and J. Swevers, "Extended Bandwidth Zero Phase Error Tracking Control of Nonminimal Phase Systems," *ASME Journal of Dynamic Systems, Measurement and Control*, Vol. 114, pp. 347–351, 1992.

[8] E. Gross, M. Tomizuka, and W. Messner, "Cancellation of Discrete Time Unstable Zeros by Feedforward Control," *ASME Journal of Dynamic Systems, Measurement and Control*, Vol. 116, pp. 33–38, 1994.

[9] E. Cooper, "Minimizing Power Dissipation in a Disk File Actuator," *IEEE Transactions on Magnetics*, Vol. 24, No. 3, pp. 2081–2091, 1988.

[10] Y. Mizoshita, S. Hasegawa, and K. Takaishi, "Vibration Minimized Access Control for Disk Drives," *IEEE Transactions on Magnetics*, Vol. 32, No. 3, pp. 1793–1798, 1996.

[11] D. Miu and S. Bhat, "Minimum Power and Minimum Jerk Position Control and Its Application in Computer Disk Drives," *IEEE Transactions on Magnetics*, Vol. 27, No. 6, pp. 4471–4475, 1991.

[12] S. Bhat and D. Miu, "Point-to-Point Positioning of Flexible Structures Using a Time Domain LQ Smoothness Constraint," *ASME Journal of Dynamic Systems, Measurement and Control*, Vol. 114, pp. 416–421, 1992.

[13] H. Yamaura, H. Ono, and M. Nishimura, "Vibrationless Acceleration Control of Positioning Mechanism and Its Application to Hard-Disk Drives," in *Proceedings of JSME International Conference on Advanced Mechatronics*, Tokyo, Japan, pp. 25–30, 1989.

[14] W. N. Patten, H. C. Wu, and L. White, "A Minimum Time Seek Controller for a Disk Drive," *IEEE Transactions on Magnetics*, Vol. 31, No. 3, pp. 2380–2387, 1995.

[15] K. S. Ananthanarayanan, "Third-Order Theory and Bang-Bang Control of Voice Coil Actuators," *IEEE Transactions on Magnetics*, Vol. 18, No. 3, pp. 888–892, 1982.

[16] R. K. Oswald, "Design of a Disk File Head-Positioning Servo," *IBM Journal of Research and Development*, pp. 506–512, 1974.

[17] R. D. Commander and J. R. Taylor, "Servo Design for an Eight-Inch Disk File," *IBM Disk Storage Technology*, pp. 89–96, 1980.

[18] S. Shimonou, H. Inada, Y. Sato, and K. Morita, "A Servo Simulator for Disk Head Positioning Servo Design," in *Proceedings of the Annual Symposium on Incremental Motion Control System Devices*, pp. 337–348, 1980.

[19] J. W. Espy, "Techniques of Position Control of High Performance, High Capacity Disk File Systems," in *Proceedings of the Annual Symposium on Incremental Motion Control System Devices*, Vol. 12, pp. 177–210, 1983.

[20] M. L. Workman, *Adaptive Proximate Time-Optimal Servomechanism*, Ph.D. Dissertation, Stanford University, 1987.

[21] M. C. Stich,"Digital Servo Algorithm for Disk Actuator Control," in *Proceedings of the Conference on Applied Motion Control*, pp. 35–41, 1987.

[22] T. H. Lee, T. S. Low, A. Al-Mamun, "DSP-Based Seek Controller for Disk Drive Servomechanism," *IEEE Transactions on Magnetics*, Vol. 29, No. 6, pp. 4071–4073, 1993.

[23] L. Y. Pao and G. Franklin, "The Robustness of a Proximate Time-Optimal Controller," *IEEE Transactions on Automatic Control*, Vol. 39, No. 9, pp. 1963–1966, 1994.

[24] T. Yamaguchi, K. Shishida, S. Tohyama, and H. Hirai, "Mode Switching Control Design with Initial Value Compensation and Its Application to Head Positioning Control on Magnetic Disk Drives," *IEEE Transactions on Industrial Electronics*, Vol. 43, No. 1, pp. 65–73, February 1996.

[25] T. Yamaguchi, Y. Soyama, H. Hosokawa, K. Tsuneta, and H. Hirai, "Improvement of Settling Response of Disk Drive Head Positioning Servo Using Mode Switching Control with Initial Value Compensation," *IEEE Transactions on Magnetics*, Vol. 32, No. 3, pp. 1767–1772, 1996.

[26] J. Ishikawa, Y. Yanagita, T. Hattori, and M. Hashimoto, "Head Positioning Control for Low Sampling Rate Systems Based on Two Degrees-of-Freedom Control," *IEEE Transactionson Magnetics*, Vol. 32, No. 3, pp. 1787–1792, 1996.

[27] J. Paxman and G. Vinnicombe, "Optimal Transfer Schemes For Switching Controllers," in *Proceedings of the 39th IEEE Conference on Decision and Control*, pp. 1093–1098, 2000.

[28] R. Oboe and M. Federico, "Initial Value Compensation Applied to Disturbance Observer-Based Servo Control in HDD," in *Proceedings of the 2002 IEEE Conference on Advanced Motion Control*, pp. 34–39, 2002.

[29] J. Yu, "Robust Initial Value Compensation Design," in *Proceedings of the 2004 ASME ISPS*, 2004.

[30] M. Johansson, "Optimal Initial Value Compensation for Fast Settling Times in Mode-Switching Control Systems," in *Proceedings of the 39th IEEE Conference on Decision and Control*, pp. 5137–5142, 2000.

[31] S. Nakagawa and T. Yamaguchi, "Settling Control System Design to Reduce Residual Vibrations During Track Seek in Magnetic Disk Drives," *Transactions of the Institute of Electrical Engineers of Japan: Part D*, Vol. 123-3, pp. 238–246, 2003 (in Japanese).

[32] T. Yamaguchi and H. Hirai, "Control of Transient Response on Servo System using Mode-Switching Control and Its Application to Magnetic Disk Drives," *Control Engineering Practice*, Vol. 6, pp. 1117–1123, 1998.

[33] Y. Li, Y. Sun, C. Smith, L. Guo, and W. Guo, "Optimization of Initial Value Compensation for Settle Control in Hard Disk Drives, in *Proceedings of the Asia Pacific Magnetic Recording Conference*, SV01-01/02, 2004.

[34] N. Hirose, M. Kawafuku, M. Iwasaki, and H. Hirai, "Residual Vibration Suppression Using Initial Value Compensation for Repetitive Positioning," *Transactions of the Institute of Electrical Engineers of Japan: Part D*, Vol. 125-1, pp. 76–83, 2005 (in Japanese).

[35] L. Yang and M. Tomizuka, "Short Seeking by Multi-rate Digital Controllers for Computation Saving with Initial Value Adjustment," in *Proceedings of the 2004 IEEE Conference on Decision and Control*, pp. 521–526, 2004.

[36] T. Izumi, A. Kojima, and S. Ishijima, "Improving Transient Responses with Control Law," *Transactions of the Institute of Systems, Control, and Information Engineers*, Vol. 13, No. 11, pp. 511–518, 2000 (in Japanese).

[37] T. Asai, "A General Synthesis Framework to Attenuate Distrubance Responses due to Switching: Synthesis of Reduced Order Compensators," *Transactions of the Society of Instrument and Control Engineers*, Vol. 42, No. 7, pp. 775–782, 2006 (in Japanese).

[38] T. Yamaguchi, H. Numasato, and H. Hirai, "A Mode-Switching Control for Motion Control and Its Application to Disk Drives: Design of Optimal Mode-Switching Conditions," *IEEE/ASME Transactions on Mechatronics*, Vol. 3, No. 3, pp. 202–209, 1998.

[39] A. Okuyama, T. Yamaguchi, K. Shishida, and T. Horiguchi, "Reduction of Acoustic Noise in a Head-Positioning Servo System of a Magnetic Disk Drive by Using Initial Value Compensation and Determining Optimal Mode Switching Conditions," *Transactions of the Japan Society of Mechanical Engineers: Part C*, Vol. 68-671, pp. 2016–2022, 2002 (in Japanese).

[40] G. F. Franklin, J. D. Powell, and M. Workman, *Digital Control of Dynamic Systems*, Addison Wesley Longman, 1998.

[41] T. Chen and L. Qiu, "H_∞ Design of General Multi-rate Sampled-Data Control Systems," *Automatica*, Vol. 30, No. 7, pp. 1139–1152, 1994 .

[42] W. W. Chiang, "Multirate State-Space Digital Controller for Sector Servo Systems," in *Proceedings of the 29th IEEE Conference on Decision and Control*, pp. 1902–1907, 1990.

[43] D. T. Phan, "The Design and Modeling of Multirate Digital Control Systems for Disk Drive Applications," in *Proceedings of the 1993 Asia-Pacific Workshop on Advanced Motion Control*, pp. 189–205, 1993.

[44] T. Hara and M. Tomizuka, "Multi-Rate Controller for Hard Disk Drives with Redesign of State Estimator," in *Proceedings of the American Control Conference*, Vol. 5, pp. 3033–3037, 1998.

Chapter 4

Ultra-Fast Motion Control

Mitsuo Hirata

Utsunomiya University

Hiroshi Fujimoto

The University of Tokyo

4.1 Vibration-Minimized Trajectory Design

In this section, reference trajectory design techniques that balance the trade-off between minimum positioning-time and smoothness of the control input are presented. The proposed methodology prevents excitation of the high frequency resonant modes which result in residual vibrations, and the effectiveness is demonstrated via application to track-seeking control in Hard Disk Drives (HDDs).

4.1.1 Introduction

In control systems, it is generally difficult to respond to the reference input quickly using only feedback control, as feedback control action occurs only when a tracking error between the reference input and the output is observed.

In such cases, the use of a *feedforward input* in addition to the feedback control is effective.

For rigid body systems, it is well-known that the *minimum time control* using maximum deceleration after a maximum acceleration (or so-called "bang-bang" input) achieves the minimum time positioning. However, all mechanical systems have resonant modes at high frequencies. As such, a rapid acceleration/deceleration input profile excites these resonant modes and the corresponding residual vibrations induced may increase the total positioning time. On the other hand, a smooth input does not excite the resonant modes, but might be unable to achieve the desired short positioning time. In view of these issues, the feedfoward input has to be designed considering the trade-off between the positioning time and smoothness of the input. When the positioning time is very short, the control input can only be updated several times during positioning, and it is better to handle such a control problem in the sampled-data framework.

In this section, a feedforward input design method which can cope with the above-mentioned problems based on *Final State Control* (FSC) is introduced [1].

4.1.2 Final State Control (FSC) Theory

FSC is a method to obtain a profile for the feedforward input that drives an initial state to a final state in N sampling steps [2]. In fact, such a problem can be formulated into a classical optimal control problem with two boundary conditions [3]. If the control input minimum norm problem is assumed in the discrete-time domain, it can be easily solved with mathematical methods in the current literature [1][2].

The state-space representation of an mth-order discrete-time controllable system is given by

$$x[k+1] = Ax[k] + Bu[k]. \tag{4.1}$$

For this system, we would like to obtain a control input $u[k]$ that drives the initial state $x[0]$ to the final-state $x[N]$ in $N \geq m$ sampling steps. Since such an input is generally not unique, and the cost function to be minimized is defined as

$$J = U^T QU, \quad Q > 0, \tag{4.2}$$

where

$$U := \begin{pmatrix} u[0], & u[1], & \dots & u[N-1] \end{pmatrix}^T. \tag{4.3}$$

By calculating (4.1) from $k = 0$ to N recursively, we have

$$x[N] - A^N x[0] = \Sigma U, \tag{4.4}$$

where

$$\Sigma = [\ A^{N-1}B, \quad A^{N-2}B, \quad \ldots \quad B \ ]. \tag{4.5}$$

It follows from the controllability of (A, B) and $N \geq m$ that Σ is a matrix of full row rank, and it can be shown that a control input U which satisfies (4.4) always exists. In order to obtain a minimum solution of (4.2) subjected to (4.4), a *Lagrange multiplier* 2λ is included and the cost function J becomes

$$J = U^T Q U + 2\lambda(X - \Sigma U), \tag{4.6}$$

where $X := x[N] - A^N x[0]$. It follows from $Q > 0$ that there is a unique solution, and the minimum solution U must satisfy

$$\frac{\partial J}{\partial U} = 2QU - 2\Sigma^T \lambda^T = 0. \tag{4.7}$$

Since $Q > 0$ and Q^{-1} exists, we have

$$U = Q^{-1}\Sigma^T \lambda^T. \tag{4.8}$$

Furthermore, it follows from $X - \Sigma U = 0$ and (4.8) that $X - \Sigma Q^{-1}\Sigma^T \lambda^T = 0$ is satisfied. Since Σ is a matrix of full rank by the controllability of (A, B), $|\Sigma Q^{-1}\Sigma^T| \neq 0$ is also satisfied. As such, we have

$$\lambda^T = (\Sigma Q^{-1}\Sigma^T)^{-1}X. \tag{4.9}$$

Finally, it follows from (4.8) and (4.9) that [1]

$$U = Q^{-1}\Sigma^T(\Sigma Q^{-1}\Sigma^T)^{-1}(x[N] - A^N x[0]). \tag{4.10}$$

The optimal feedforward input U can now be easily obtained using simple algebra for given $x[0]$, $x[N]$, Q, and N.

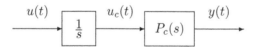

FIGURE 4.1: Augmented system for SMART trajectory.

4.1.3 Vibration Minimized Trajectory Design Based on Final State Control

In the previous chapter, the minimum jerk input is introduced as a reference input for Two-Degrees-of-Freedom (TDOF) control. Since the minimum jerk input is smooth, it is difficult to excite the high frequency resonant modes of the plant. As shown previously, one method to obtain a smooth input is

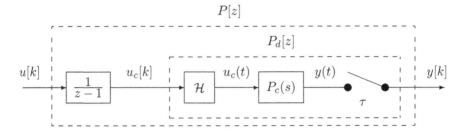

$$P[z]$$

FIGURE 4.2: Augmented system with a discrete-time integrator.

to use the augmented system of Figure 4.1. Since $u(t)$ is the derivative of the actual input $u_c(t)$, a smooth input can be obtained by minimizing the *square integral* of $u(t)$ subject to the initial and final conditions [4].

Although the minimum jerk input is a continuous-time signal, it can be applied to the discrete-time plant by sampling the continuous-time input at every instant. This method works well when the sampling period is short enough. However, if a positioning error exists, e.g., from discretization error, a problem reformulation in the discrete-time domain is necessary.

Similar to Figure 4.1, a discrete-time integrator $\frac{1}{z-1}$ is connected to the input $P_d[z]$, as shown in Figure 4.2. The state-space representation of $P_c(s)$ which is of mth-order is given by

$$\dot{x}_c(t) = A_c x_c(t) + B_c u_c(t), \quad y(t) = C_c x_c(t). \tag{4.11}$$

$P_d[z]$ is a discrete-time model of $P_c(s)$ discretized using a Zero-Order Hold (ZOH), and its state-space representation is given by

$$x_d[k+1] = A_d x_d[k] + B_d u_c[k], \quad y[k] = C_d x_d[k], \tag{4.12}$$

where $A_d := e^{A_c \tau}$, $B_d := \int_0^\tau e^{A_c t} B_c dt$, $C_d := C_c$, $x_d[k] = x_c(k\tau)$, $y[k] = y(k\tau)$, and τ is the sampling period. Since the ZOH is used, we have

$$u_c(k\tau + \theta) = u_c[k], \quad \theta \in [0, \tau). \tag{4.13}$$

The state-space representation of the augmented system is thus given by

$$x[k+1] = Ax[k] + Bu[k], \quad y[k] = Cx[k], \tag{4.14}$$

where $x[k] := [x_d^T[k], u_c^T[k]]^T$ and

$$A = \begin{bmatrix} A_d & B_d \\ 0 & 1 \end{bmatrix}, \quad B = \begin{bmatrix} 0 \\ 1 \end{bmatrix}, \quad C = \begin{bmatrix} C_d & 0 \end{bmatrix}. \tag{4.15}$$

In this case, both (A_c, B_c) and (A_d, B_d) are assumed to be controllable.

Now, a smooth input is obtained by applying the FSC method to the augmented system in (4.14) via minimizing $J = \sum_{k=0}^{N-1} u^2[k] = U^T U$ subject to both initial and final state constraints. It should be noted that the initial and final states of the augmented system have to be given by

$$x[0] = 0_{(m+1)\times 1}, \quad x[N] = \begin{bmatrix} x_d[N] \\ 0 \end{bmatrix}, \qquad (4.16)$$

so that the actual input u_c satisfies $u_c[0] = 0$ and $u_c[N] = 0$. The feedforward input obtained by this method is referred to as the *FSC input* and the trajectory generated by the FSC input is referred to as the *FSC trajectory*.

In the FSC method, a smooth input is desired to reduce the residual vibrations after positioning. However, such an input may extend beyond the total positioning time. In order to achieve both high-speed positioning and vibration suppression, we consider minimizing the *spectrum of the control input* at the frequencies where resonant modes of the plant exist.

The Fourier transform $\hat{U}_c(\omega)$ of $u_c(t)$ is defined as

$$\hat{U}_c(\omega) = \int_0^{N\tau} u_c(t)e^{-j\omega t}dt, \qquad (4.17)$$

and its gain $|\hat{U}_c(\omega)|$ is minimized at the desired frequency points $\omega_i(i = 1,\ldots,\ell)$. For this purpose, the following cost function is introduced as

$$J_w = \sum_{k=0}^{N-1} u^2[k] + \sum_{i=1}^{\ell} q_i|\hat{U}_c(\omega_i)|^2. \qquad (4.18)$$

The first term of the right-hand side of (4.18) evaluates the smoothness of the control input, and the second term evaluates the frequency components of the control input. By choosing ω_i to cover the frequencies of the resonant modes of the plant, the frequency components around these frequencies are reduced. The weighting parameter q_i is selected to be a positive real number, and a larger q_i achieves larger reduction in frequency components.

In order to obtain the closed-form of U which minimizes (4.18), the state-space representation of the discrete-time integrator is given by

$$\frac{1}{z-1} = \{1,1,1,0\} =: \{A_w, B_w, C_w, 0\}. \qquad (4.19)$$

By assuming $u_c[0] = 0$, the actual control input $u_c[k]$ can be described by

$$U_c = \Omega_w U, \qquad (4.20)$$

where

$$U_c = \begin{bmatrix} u_c[0],\ldots,u_c[N-1] \end{bmatrix}^T \qquad (4.21)$$

and

$$\Omega_w = \begin{bmatrix} 0 & 0 & \cdots & 0 \\ C_w B_w & 0 & \ddots & \vdots \\ \vdots & \ddots & \ddots & 0 \\ C_w A_w^{N-2} B_w & \cdots & C_w B_w & 0 \end{bmatrix}. \qquad (4.22)$$

Since $u_c(t)$ is the output of the ZOH, we have

$$u_c(t) = \sum_{k=0}^{N-1} P_i(t)u_c[k], \tag{4.23}$$

where

$$P_i(t) := \begin{cases} 1, & i\tau \le t < (i+1)\tau \\ 0, & t < i\tau \text{ or } t \ge (i+1)\tau \end{cases} . \tag{4.24}$$

The Fourier transform $\hat{U}_c(\omega)$ of $u_c(t)$ can be given by

$$
\begin{aligned}
\hat{U}_c(\omega) &= \int_0^{N\tau} u_c(t)e^{-j\omega t}dt \\
&= \sum_{k=0}^{N-1} \int_{k\tau}^{(k+1)\tau} u_c[k]e^{-j\omega t}dt \\
&= \frac{2\sin(\omega\tau/2)}{\omega}e^{-j\omega\tau/2} \sum_{k=0}^{N-1} u_c[k]e^{-j\omega\tau k} \\
&= \hat{U}_1(\omega)\hat{U}_2(\omega)e^{-j\omega\tau/2},
\end{aligned} \tag{4.25}
$$

where

$$\hat{U}_1(\omega) := \frac{2\sin(\omega\tau/2)}{\omega}, \quad \hat{U}_2(\omega) := \sum_{k=0}^{N-1} u_c[k]e^{-j\omega\tau k}. \tag{4.26}$$

It is easy to show that

$$\Re\left[\hat{U}_2(\omega)\right] = \sum_{k=0}^{N-1} u_c[k]\cos(k\omega\tau) = S_R(\omega)U_c, \tag{4.27}$$

$$\Im\left[\hat{U}_2(\omega)\right] = \sum_{k=0}^{N-1} u_c[k]\sin(k\omega\tau) = S_I(\omega)U_c, \tag{4.28}$$

where

$$S_R(\omega) := \left[\cos(0), \cos(\omega\tau), \ldots, \cos((N-1)\omega\tau)\right], \tag{4.29}$$

$$S_I(\omega) := \left[\sin(0), \sin(\omega\tau), \ldots, \sin((N-1)\omega\tau)\right]. \tag{4.30}$$

As such,

$$|\hat{U}_2(\omega)|^2 = U_c^T(S_R^T S_R(\omega) + S_I^T S_I(\omega))U_c. \tag{4.31}$$

From (4.31) and $U_c = \Omega_z U$, we have

$$|\hat{U}_c(\omega)|^2 = |\hat{U}_1(\omega)|^2 \cdot U^T \Omega_z^T (S_R^T S_R(\omega) + S_I^T S_I(\omega)) \Omega_z U. \qquad (4.32)$$

Finally, J_w in (4.18) can be represented as a quadratic form of U as

$$J_w = U^T Q_w U, \qquad (4.33)$$

where

$$Q_w = I_N + \sum_{i=1}^{\ell} q_i Q_U(\omega_i), \qquad (4.34)$$

$$Q_U(\omega_i) = |\hat{U}_1(\omega_i)|^2 \cdot \Omega_z^T (S_R^T S_R(\omega_i) + S_I^T S_I(\omega_i)) \Omega_z. \qquad (4.35)$$

The FSC input obtained by minimizing (4.33) is referred to as *Frequency-shaped Final State Control (FFSC) input* and the trajectory generated by the FFSC input is referred to as the *FFSC trajectory*.

4.1.4 Application to Track-Seeking Control in HDDs

In this section, the feedforward inputs are designed by using the FSC and FFSC methods. The plant is a low-end 2.5" HDD used in notebook computers, and its sampling period is 105 μs, which is considered to be relatively long. The measured frequency response of the plant is shown in Figure 4.3, and the plant has multiple resonant modes at the high frequency region above 4 kHz. In this case, we assumed that the resonant modes with varying natural frequencies are difficult to model, and the nominal model is defined by a rigid body mode as

$$P_c(s) := \frac{k_p}{s^2}, \quad y = P_c u_c, \qquad (4.36)$$

where the control input u_c is a voltage input to the Voice Coil Motor (VCM) in volts, the measurement output y is the head position in number of tracks, and $k_p = 3.688 \times 10^{10}$. The frequency response of (4.36) is shown by the solid line in Figure 4.3.

In order to calculate the FSC and FFSC inputs for P_c, the initial states are defined as zero position and zero velocity, and the final states are defined as two tracks in position and zero velocity. The final time is set at $N = 10$. To ensure that the mechanical resonant modes at high frequencies around 4 kHz are not excited, the frequency points ω_i in (4.18) are specified from 3.5 kHz to 7.5 kHz by dividing the range equally into two hundred and thirty frequency points. The weight q_i is selected as $q_i = 1 \times 10^{11}$ for all frequency points.

The obtained FSC and FFSC inputs are shown in Figure 4.4. The upper and lower figures show the time responses and their corresponding frequency spectra, respectively. From the spectrum of FFSC input, it can be observed that the spectrum from 3.5 kHz to 7.5 kHz is reduced by up to -20 dB. Since these frequencies are around the Nyquist frequency of 5 kHz, it may be difficult

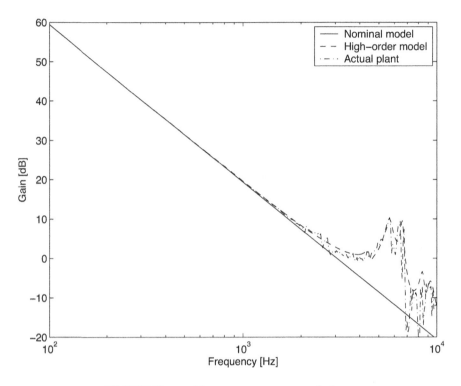

FIGURE 4.3: Frequency response of plant.

for a general digital control method to achieve such an impressive result. Moreover, the proposed method can reduce the frequency components around the Nyquist frequency *without* even employing multi-rate control techniques.

In order to show the effectiveness of the proposed method in the time domain, the output responses are simulated by using a precise model with five mechanical resonant modes denoted by the dashed lines in Figure 4.3. In this simulation, a feedback controller is not used since no disturbances are injected into the system. Figure 4.5 shows the tracking error between the output of the rigid body model (nominal model) and the precise model. From this figure, it can be verified that the residual vibrations of the proposed method during track-seeking are reduced after 1 ms.

Next, experiments were carried out. In the experiments, a feedback controller was implemented due to presence of various disturbance sources. The feedback controller was designed by the discrete-time H_∞ control method, and was implemented by a model-matching TDOF control method as shown in Figure 4.6 [1]. In this figure, K is the H_∞ controller and $u[k]$ is a feedforward input obtained by either the FSC or FFSC method. In the TDOF control method, the feedback controller works only when the output $y[k]$ of the actual plant and the output $y_m[k]$ of the model P_d differ.

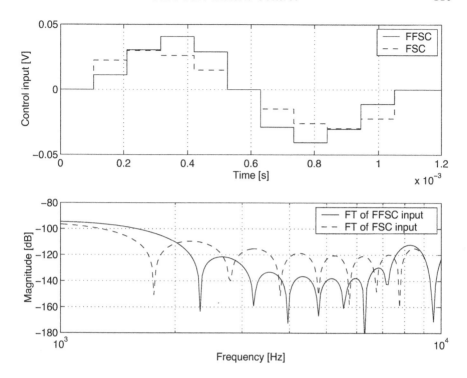

FIGURE 4.4: Control inputs $u(t)$ and their frequency spectra.

The output responses from the experiments are shown in Figure 4.7, and the magnified plots around the target track are shown in Figure 4.8. The track-seeking experiments were repeated fourteen times, and the average and the envelope of $\pm 3\sigma$ of the responses are shown by the thick and thin lines, respectively. It is worth noting that the head position has not arrived at the target track as external forces caused by the flexible cable and/or gravity are not compensated in these experiments.

Upon closer inspection of the responses after the track-seeking, the FFSC method achieves smaller residual vibrations as compared to the FSC method. From the envelope of $\pm 3\sigma$, it can be seen that the fluctuation of the response is also reduced by the FFSC method. The power spectrum densities of the tracking errors were calculated and are shown in Figure 4.9. It can be seen that the magnitude of errors beyond 3.5 kHz is reduced by the FFSC method.

From the experimental results, we can conclude that the FFSC method is effective for fast and accurate positioning systems where the plant has mechanical resonant modes at high frequencies, such as the track-seeking control problem in HDDs.

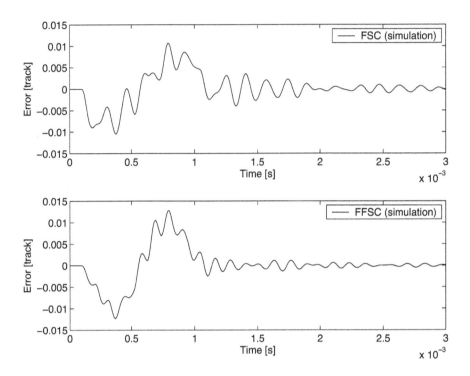

FIGURE 4.5: Displacement profile for two-track seek.

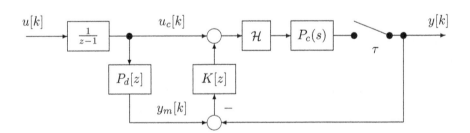

FIGURE 4.6: Block diagram of TDOF system for implementation of proposed feedforward input.

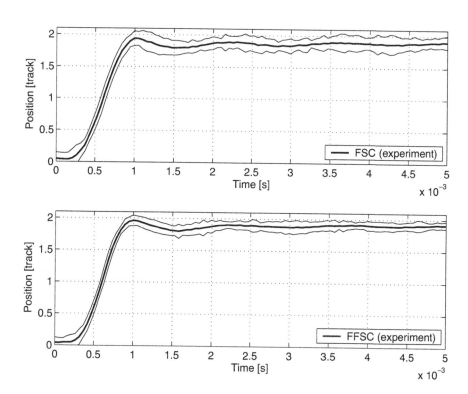

FIGURE 4.7: Head positions for two-track seek control.

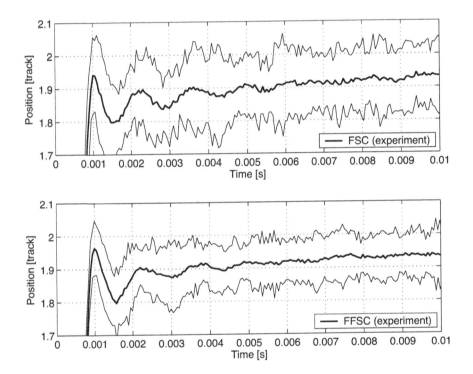

FIGURE 4.8: Head positions for two-track seek control (magnified).

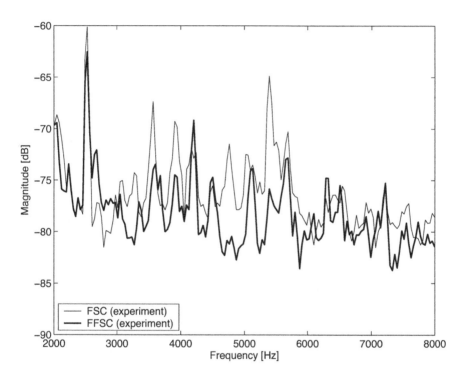

FIGURE 4.9: Power spectrum densities of tracking error.

4.2 Perfect Tracking Control (PTC)

In this section, the concepts of Perfect-Tracking Control (PTC) and several multi-rate feedforward PTC techniques that enable the plant output to perfectly track a desired trajectory are presented. The proposed control techniques are applied to the track-seeking mode in HDDs, and the effectiveness of PTC is verified through simulations and experimental implementations in actual HDDs.

4.2.1 Introduction

There are generally two control modes, namely the track-seeking and track-following modes, in head-positioning control systems for HDDs. In the track-seeking mode, the feedforward performance is important because the head should be actuated to the desired track as quickly as possible. In the track-following mode, the head should be positioned precisely on the desired track in the presence of vibrations generated by the disk rotation and disturbances while the Read/Write (R/W) processes are being carried out. The disturbance rejection performance is important because the head must be positioned accurately on the desired track. During a long-span seek where the seeking distance is comparatively long, high-speed seeking is achieved by Mode Switching Control (MSC) [5]. However, single mode controllers based on TDOF control are commonly used during a short-span seek instead, as the MSC might generate an undesirable transient response on some occasions [6].

Digital TDOF controllers generally have two samplers for the reference signal $r(t)$ and the output $y(t)$, as well as one hold for the input $u(t)$. As such, there exist three time periods T_r, T_y, and T_u, which represent the periods of $r(t)$, $y(t)$, and $u(t)$, respectively. The control period T_u is generally decided by the speed of the actuator, Digital-to-Analog (D/A) converter, or the computational speed of the CPU, etc. On the other hand, the output sampling period T_y is determined by the speed of the sensor or the Analog-to-Digital (A/D) converter, etc.

In the case of HDDs, position error is detected by the discrete servo signals embedded in the disks. Therefore, the output sampling period T_y is restricted by the number of these signals and the rotational frequency of the spindle motor. However, it is possible to set the control period T_u to be shorter than T_y because current digital signal processors are of high computational performance. With this technology, the controller can be implemented as a *multi-rate control system* with a hardware constraint of $T_u < T_y$. Currently, many multi-rate controllers have been proposed for both track-seeking and track-following modes in the current literature [7]–[20].

The term *Perfect Tracking Control* (PTC) was originally defined in [21], where the plant output perfectly tracks the desired trajectory with zero tracking error at *every* sampling point. However, it is not possible to implement PTC in conventional single-rate control systems because the discrete-time plant with a ZOH has unstable zeros when the relative degree of the plant is greater than two, even in the ideal case where the plant has no modeling error and disturbances [22].

On the other hand, the PTC method using multi-rate *feedforward control* instead of the ZOH are developed in [23] and [24]. In this PTC, the tracking error of the state of the plant becomes exactly zero at every sampling period of reference input for a nominal plant without disturbance. By combining the proposed feedforward controller with a robust feedback controller, high tracking performance can be preserved even if the actual plant has modeling error and disturbances.

The development of PTC leads to its successful application to the track-seeking mode of HDDs where the plant is modeled as a rigid body with time delay [13]. The multi-rate feedforward control is suitable with the constraint of a longer T_y. With the increased demand for faster user data access during short-span seeks, it is hence imperative for the primary mechanical resonant mode to be considered on top of the rigid body mode.

In [25], a novel desired trajectory for the head position is proposed for the PTC algorithm in order to minimize the residual vibrations due to the primary vibration mode. However, only the rigid body mode of second-order was considered as the nominal model when designing the feedforward controller. In this section, the PTC is designed based on the nominal model composed of the rigid body and primary vibration mode. To apply PTC to a fourth-order model, the selection of the state variable is extremely important as this algorithm guarantees perfect tracking for all the state variables at the sampling points of the reference trajectory. This section introduces a method based on a modified controllable canonical realization to obtain a smooth control input and better inter-sample response [26][27].

4.2.2 PTC Theory

According to [24], PTC is achieved when the feedforward control input is changed n times within one sampling period of the reference signal T_r, as shown in Figure 4.10, where n is the order of nominal plant. The output sampling period T_y can be determined independently according to constraints imposed by the hardware.

Consider the continuous-time n^{th} order plant described by

$$\dot{\mathbf{x}}(t) = \mathbf{A}_c\mathbf{x}(t) + \mathbf{b}_c u(t), \quad y(t) = \mathbf{c}_c\mathbf{x}(t). \tag{4.37}$$

The discrete-time state equation discretized using the shorter sampling period T_u becomes

$$\mathbf{x}[k+1] = \mathbf{A}_s\mathbf{x}[k] + \mathbf{b}_s u[k], \tag{4.38}$$

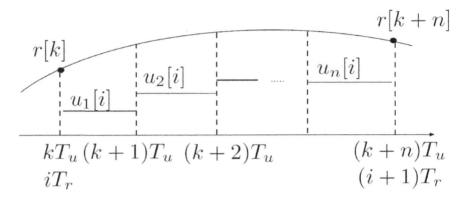

FIGURE 4.10: Multi-rate hold.

where $\mathbf{x}[k] = \mathbf{x}(kT_u)$ and

$$\mathbf{A}_s := e^{\mathbf{A}_c T_u}, \quad \mathbf{b}_s := \int_0^{T_u} e^{\mathbf{A}_c \tau} \mathbf{b}_c d\tau. \tag{4.39}$$

By calculating the state transition from $t = iT_r = kT_u$ to $t = (i+1)T_r = (k+n)T_u$, as shown in Figure 4.10, the discrete-time plant $P[z]$ can be represented by

$$\mathbf{x}[i+1] = \mathbf{A}\mathbf{x}[i] + \mathbf{B}\mathbf{u}[i], \quad y[i] = \mathbf{c}\mathbf{x}[i], \tag{4.40}$$

where $\mathbf{x}[i] = \mathbf{x}(iT_r)$, $z := e^{sT_r}$, and the multi-rate input vector $\mathbf{u}$ is defined in the lifting form as

$$\begin{aligned}
\mathbf{u}[i] &:= [u_1[i], \cdots, u_n[i]]^T \\
&= [u(kT_u), \cdots, u((k+n-1)T_u)]^T,
\end{aligned} \tag{4.41}$$

where $u_j[i]$ is the inter-sample control input shown in Figure 4.10. The coefficients of (4.40) are given by

$$\mathbf{A} = \mathbf{A}_s^n, \quad \mathbf{B} = [\mathbf{A}_s^{n-1}\mathbf{b}_s, \mathbf{A}_s^{n-2}\mathbf{b}_s, \cdots, \mathbf{A}_s\mathbf{b}_s, \mathbf{b}_s], \tag{4.42}$$

$$\mathbf{c} = \mathbf{c}_c. \tag{4.43}$$

From (4.40), the transfer function from the plant state $\mathbf{x}[i+1] \in \mathbf{R}^n$ to the multi-rate input $\mathbf{u}[i] \in \mathbf{R}^n$ and output $y[i]$ can be derived as

$$\mathbf{u}[i] = \mathbf{B}^{-1}(\mathbf{I} - z^{-1}\mathbf{A})\mathbf{x}[i+1] \tag{4.44}$$

$$= \left[\begin{array}{c|c} \mathbf{O} & -\mathbf{A} \\ \hline \mathbf{B}^{-1} & \mathbf{B}^{-1} \end{array}\right] \mathbf{x}[i+1]. \tag{4.45}$$

From the definition in (4.42), the non-singularity of matrix $\mathbf{B}$ is assured for a controllable plant. Moreover, (4.45) indicates that all poles of the transfer function (4.44) are located at the origin of the z-plane. Hence, (4.44) is a stable inverse system. If the plant has no modeling error, the feedforward control input

$$\mathbf{u}_{ff}[i] = \mathbf{B}^{-1}(\mathbf{I} - z^{-1}\mathbf{A})\mathbf{r}[i], \tag{4.46}$$

where $\mathbf{r}[i](:= \mathbf{x}_d[i+1])$ is the previewed desired trajectory of the plant state which can guarantee perfect tracking at every sampling period T_r [24]. When the tracking error $\mathbf{e} = y[i] - y_0[i]$ arises due to the presence of modeling error or disturbances, it can be attenuated by the robust feedback controller $\mathbf{C}_2[z]$ designed as

$$\mathbf{u}[i] = \mathbf{u}_{ff}[i] + \mathbf{C}_2[z](\mathbf{y}[i] - \mathbf{y}_0[i]), \tag{4.47}$$

where $\mathbf{y}_0[i]$ is the nominal output and is the ideal output signal when perfect tracking is achieved [24].

4.2.3 Vibration Suppression Using PTC

In this section, several PTC techniques that minimize vibrations of the plant caused by excitation of high frequency resonant modes are presented.

4.2.3.1 With MPVT

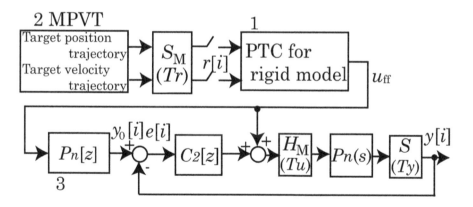

FIGURE 4.11: Vibration suppression PTC by MPVT.

Figure 4.11 shows the vibration suppression PTC technique by *Minimizing Primary Vibration Trajectory* (MPVT) which was proposed in [25]. In Figure 4.11, block 1 generates the acceleration feedforward input which guarantees perfect tracking for the rigid body mode. This rigid body model can be represented by a double integrator as

$$\mathbf{A}_c = \begin{bmatrix} 0 & 1 \\ 0 & 0 \end{bmatrix}, \mathbf{b}_c = \begin{bmatrix} 0 \\ K_p \end{bmatrix}, \mathbf{c}_c = [1, 0], \tag{4.48}$$

where K_p is the plant gain.

Block 2 generates the seeking trajectory that minimizes the residual vibrations caused by the primary vibration mode of the actuator. This scheme is known as the MPVT, where the head-position trajectory is obtained from the step response of the transfer function given by

$$G_m(s) = \frac{\frac{1}{w_r^2}s^2 + \frac{2\zeta_r}{w_r}s + 1}{\prod_{i=1}^{n}(\tau_i s + 1)},$$

(4.49)

where τ_i is a time constant of the response, w_r is the angular frequency of the primary mechanical vibration, and ζ_r is the damping ratio. The head-velocity trajectory is given by the impulse response of (4.49). This transfer function is designed to include the band-stop characteristics using the numerator polynomials.

As the PTC is designed for the rigid body model, block 3 is included to reduce the effects of modeling error of the target trajectory on the feedback controller $C_2[z]$. The plant model consisting of the rigid body and primary vibration mode is given as

$$P_n(s) = K_p \left(\frac{1}{s^2} + \frac{a_r}{s^2 + 2\zeta_r w_r s + w_r^2} \right).$$

(4.50)

4.2.3.2 With Parallel Realization

In MPVT, the feedforward PTC is designed using the second-order rigid body mode as the nominal plant. As such, it is essential to design a desired trajectory which does not excite the primary vibration mode. In this section, PTC is designed based on the fourth-order nominal model with the rigid body and primary vibration modes in order to track the desired trajectory more accurately.

A state-space realization of the plant cannot be uniquely determined from the transfer function depicted in (4.50). One straightforward approach is based on the *parallel realization* of rigid body mode and primary vibration mode and the corresponding state-space model is given as

$$\mathbf{A}_c = \begin{bmatrix} 0 & 1 & 0 & 0 \\ 0 & 0 & 0 & 0 \\ 0 & 0 & 0 & 1 \\ 0 & 0 & -w_r^2 & -2\zeta_r w_r \end{bmatrix}, \mathbf{b}_c = \begin{bmatrix} 0 \\ K_p \\ 0 \\ K_p a_r \end{bmatrix},$$

(4.51)

$$\mathbf{c}_c = \begin{bmatrix} 1 & 0 & 1 & 0 \end{bmatrix},$$

(4.52)

where x_1 and x_2 are the position and velocity of rigid body mode, and x_3 and x_4 are the position and velocity of the primary vibration mode, respectively. This formulation of PTC is shown in Figure 4.12.

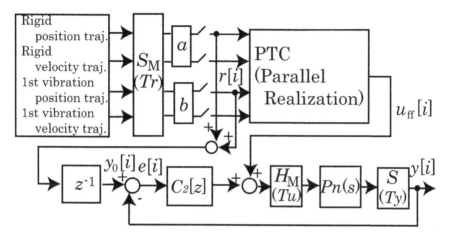

FIGURE 4.12: Vibration suppression PTC by parallel realization.

When the desired position of each mode is given as $x_{1d}(s) = \frac{a}{s(\tau s+1)^m}$ and $x_{3d}(s) = \frac{b}{s(\tau s+1)^m}$, the desired output trajectory becomes $\frac{a+b}{s(\tau s+1)^m}$. As such, the ratio of each mode can be changed under the constraint of $a+b=1$, and the desired velocities are given by the derivatives of each mode.

4.2.3.3 With Modified Controllable Canonical Realization

In the modified controllable canonical realization mode, the PTC can be designed based on the following model

$$P_n(s) = K_p \frac{(1+a_r)s^2 + 2\zeta_r \omega_r s + \omega_r^2}{s^4 + 2\zeta_r \omega_r s^3 + \omega_r^2 s^2} \tag{4.53}$$

which is converted from (4.50) and is shown in Figure 4.13. The modified controllable canonical realization is represented as

$$\mathbf{A}_c = \begin{bmatrix} 0 & 1 & 0 & 0 \\ 0 & 0 & 1 & 0 \\ 0 & 0 & 0 & 1 \\ 0 & 0 & -\omega_r^2 & -2\zeta_r \omega_r \end{bmatrix}, \mathbf{b}_c = \begin{bmatrix} 0 \\ 0 \\ 0 \\ K_p \omega_r^2 \end{bmatrix}, \tag{4.54}$$

$$\mathbf{c}_c = \begin{bmatrix} 1 & \frac{2\zeta_r}{\omega_r} & \frac{1+a_r}{\omega_r^2} & 0 \end{bmatrix}. \tag{4.55}$$

While $\mathbf{b}_c = [0,0,0,1]^T$ in the original controllable canonical form, it has a gain $K_p \omega_r^2$ in the modified realization form.

In PTC, the desired trajectory should be given to all the state variables of the plant. From (4.54), the desired state is given as

$$\mathbf{x}_d(t) = \left[z_d(t), z_d^{(1)}(t), z_d^{(2)}(t), z_d^{(3)}(t)\right]^T, \tag{4.56}$$

where $z_d(t)$ is the desired trajectory of $x_1(t)$ and $z_d^{(n)}$ is the nth-order derivative. Note that $z(t)$ is not the head position since $y(t) \neq x_1(t)$. To make things clearer, the transfer function from u to $z = x_1$ is calculated as

$$\frac{z}{u} = K_p \frac{\omega_r^2}{s^4 + 2\zeta_r\omega_r s^3 + \omega_r^2 s^2}. \tag{4.57}$$

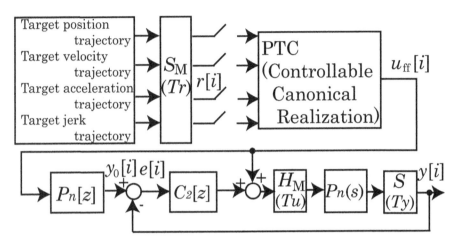

FIGURE 4.13: Vibration suppression PTC with canonical form.

The frequency responses of the transfer functions $\frac{y(s)}{u(s)}$ and $\frac{z(s)}{u(s)}$ are shown in Figure 4.14. From Figure 4.14, it can be observed that the two responses match very well in the low frequency range. This is because the first element of the coefficient vector $\mathbf{c}_c$ is normalized to unity in the modified controllable canonical realization depicted in (4.55). As such, the state variable z is position in m. In this section, this variable is called the "virtual position." Although this variable does not refer to the head position directly, vibration suppression PTC can be realized by injecting the desired trajectory to the virtual position.

4.2.4 Simulations and Experiments

In this section, the effectiveness of the proposed PTC techniques presented is verified with simulations and experiments on actual HDDs. Analyses and comparisons among the various PTC techniques are also presented.

4.2.4.1 Simulations Using Nominal Model

One-track seek simulations are performed using the vibration suppression PTCs in Section 4.2.3 on the nominal plant model as depicted in (4.50). We set the tolerances for overshoot and undershoot to be within $\pm 5\%$ of a track.

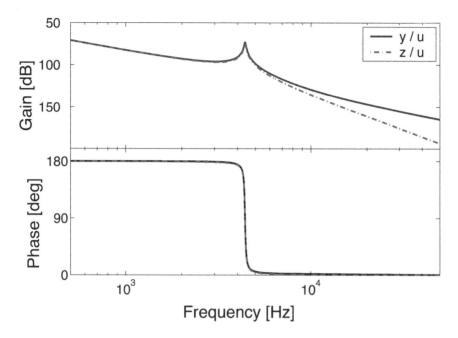

FIGURE 4.14: Frequency responses of nominal plant.

TABLE 4.1: Control and Sampling Periods

Plant model for PTC	T_r (μs)	T_u (μs)	T_y (μs)
Rigid	38.4	19.2	38.4
Rigid + Vibration	76.7	19.2	38.4

The Track Pitch (TP) is set to 0.504 μm, and the nominal parameters of the plant are $\omega_r = 2\pi 4400$ rad/s, $\zeta_r = 0.01$, $a_r = -1.2$, and $K_p = 2.77 \times 10^2$.

Table 4.1 shows the trajectory sampling period T_r, output sampling period T_y, and control period T_u for each method. In MPVT, the input multiplicity $N := \frac{T_r}{T_u}$ is two since the PTC utilizes the rigid body model. On the other hand, $N = 4$ is used in the parallel realization and canonical form methods as a fourth order model is assumed in the PTC design. In order to make a fair comparison, T_u is set to be the same for both cases and this makes T_r of the canonical form method twice as long as that of the MPVT. The desired trajectory is given as

$$\frac{1}{s(\tau s + 1)^8},\tag{4.58}$$

where the time constant τ is a tuning parameter used to change the seek time.

Figure 4.15(a) shows that MPVT works well for $\tau \geq 15$ μs. However, it excites the primary vibration mode for $\tau = 10$ μs as the feedforward PTC does not consider the primary vibration mode.

The simulation of parallel realization is shown in Figure 4.15(b). The contribution ratio of the rigid body mode is varied at 100%, 50%, and 0%, i.e., $a = 1$, 0.5, and 0 in Figure 4.12. In all cases, perfect tracking of the reference trajectory is achieved at every sampling instant T_r. However, inter-sample vibration is generated when $a = 0.5$ and $a = 0$ since the trajectory is also delivered to the primary vibration mode. As such, only the case with $a = 1$ and $b = 0$ is considered in the following simulations.

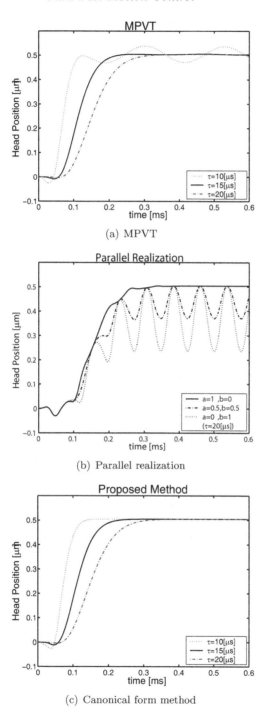

(a) MPVT

(b) Parallel realization

(c) Canonical form method

FIGURE 4.15: Simulation of nominal model.

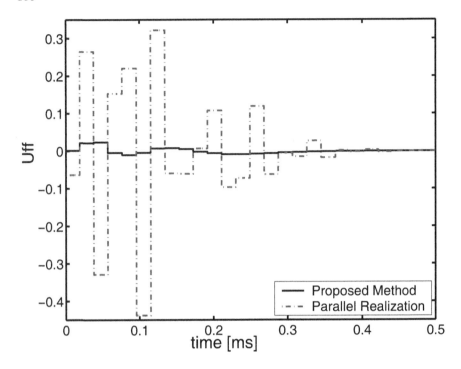

FIGURE 4.16: Control input.

Figure 4.15(c) shows the results using the modified controllable canonical realization method. From Figure 4.15(c), it can be observed that this method retains the vibration suppression characteristics even in the case when $\tau = 10$ μs and the MPVT loses its suppression capability. Moreover, the intersample response of the canonical form method is much smoother than that using parallel realization. The canonical form method also has a smooth control input as it tracks the smooth desired trajectory of position, velocity, acceleration, and jerk of the virtual plant. This is preferred as compared to the parallel realization method where the control input oscillates back and forth during one sampling period as it attempts to drive the position and velocity of the vibration mode to zero during the sampling instants as can be seen in Figure 4.16.

4.2.4.2 Experiments on HDDs

For our experiments, $y_0[k]$ as shown in Figure 4.13 is calculated at every T_y using

$$P_n[z] = Z\left[\frac{1 - e^{-T_y s}}{s}P_n(s)e^{-Ls}\right], \tag{4.59}$$

$$y_0[k] = P_n[z]u_{ff}[k], \tag{4.60}$$

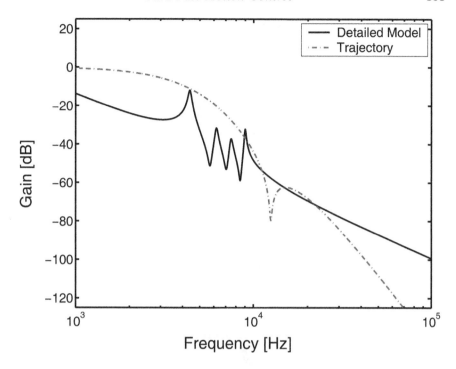

FIGURE 4.17: Frequency responses of detailed model and modified trajectory.

where e^{-Ls} with $L = 20$ μs is the equivalent time delay considering calculation delay and power amplifier, etc., for reducing the modeling error.

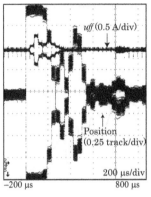

(a) Rigid PTC

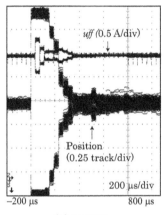

(b) MPVT

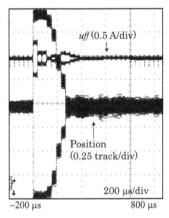

(c) Canonical form method

FIGURE 4.18: Experimental results (envelop).

TABLE 4.2: Experimental Seek Time

Rigid PTC	MPVT	Canonical form
—	0.347 ms	0.291 ms

To suppress the high order vibration modes, the modified trajectory is introduced using the canonical form method as

$$\frac{\frac{1}{\omega_h^2}s^2 + \frac{2\zeta_h}{\omega_h}s + 1}{s(\tau s + 1)^8}, \tag{4.61}$$

where ζ_h and ω_h are tuning parameters used to suppress the high order vibration modes. The frequency responses of the high order plant model considered and the corresponding velocity trajectory are shown in Figure 4.17 when $\omega_h = 11.5$ kHz.

Figures 4.18(a)–4.18(c) show the envelope from one thousand consecutive alternative seeking experiments and the target track is displayed as the center line of the vertical axis. The response of the original PTC is shown in Figure 4.18(a), where the vibration modes are not considered in the feedforward controller. This method is called the "rigid PTC" [13]. The results of the canonical form method in Figure 4.18(c) are compared with those using the rigid PTC and MPVT. While the rigid PTC excites the primary vibration mode and the MPVT has residual vibrations, the canonical form method can suppress all the vibrations very well. The advantage of the proposed canonical form method has been verified with extensive experiments.

Table 4.2 shows the average seek time obtained from the experiments. In this table, the seek time is defined as the time when the head position reaches the $\pm10\%$ band of the target track. For the rigid PTC, the seek time was not measurable since it crosses the band multiple times. It can also be observed that the proposed canonical form method has the fastest seek performance from these experiments.

In this section, the novel vibration suppression PTC with modified controllable canonical realization is presented. This method was applied to the short span seek in HDDs. While the original rigid PTC excites the mechanical resonant modes when the seek time is very short, the proposed canonical form method can suppress the primary vibration mode very well. The canonical form method also actively controls the primary vibration mode via multi-rate feedforward control, as compared to notch-type trajectory suppression using the MPVT. In conclusion, the proposed methodology has many advantages over MPVT by introducing a modified trajectory to reduce the high order vibration modes in experiments.

134

Bibliography

[1] M. Hirata, T. Hasegawa, and K. Nonami, "Short Track-Seeking Control of Hard Disk Drives by Using Final-State Control," *Transactions of the IEEJ*, Vol. 125, No. 5, pp. 524–529, 2005 (in Japanese).

[2] T. Totani and H. Nishimura, "Final-State Control Using Compensation Input," *Transactions of the SICE*, Vol. 30, No. 3, pp. 253–260, 1994.

[3] F. L. Lewis and V. L. Syrmos, *Optimal Control*, Wiley-Interscience, 1995.

[4] Y. Mizoshita, S. Hasegawa, and K. Takaishi, "Vibration Minimized Access Control for Disk Drives," *IEEE Transactions on Magnetics*, Vol. 32, No. 3, pp. 1793–1798, 1996.

[5] T. Yamaguchi, H. Numasato, and H. Hirai, "A Mode-Switching Control for Motion Control and Its Application to Disk Drives: Design of Optimal Mode-Switching Conditions," *IEEE/ASME Transactions on Mechatronics*, Vol. 3, No. 3, pp. 202–209, 1998.

[6] L. Yi and M. Tomizuka, "Two-Degrees-of-Freedom Control with Robust Feedback Control for Hard Disk Servo Systems," *IEEE/ASME Transactions on Mechatronics*, Vol. 4, No. 1, pp. 17–24, 1999.

[7] W-. W. Chiang, "Multirate State-Space Digital Controller for Sector Servo Systems," in *Proceedings of the IEEE CDC*, pp. 1902–1907, 1990.

[8] A. M. Phillips and M. Tomizuka, "Multirate Estimation and Control Under Time-Varying Data Sampling with Application to Information Storage Devices," in *Proceedings of the ACC*, pp. 4151–4155, 1995.

[9] P. A. Weaver and R. M. Ehrlich, "The Use of Multirate Notch Filters in Embedded-Servo Disk Drives," in *Proceedings of the ACC*, pp. 4156–4160, June 1995.

[10] M. Kobayashi, T. Yamaguchi, I. Oshimi, Y. Soyama, Y. Hata, and H. Hirai, "Multirate Zero Phase Error Feedforward Control For Magnetic Disk Drives," in *Proceedings of the JSME, IIP '98*, pp. 21–22, August 1998 (in Japanese).

[11] S. Takakura, "Design of the Tracking System Using N-Delay Two-Degrees-of-Freedom Control and Its Application to Hard Disk Drives," in *Proceedings of the IEEE CCA*, pp. 170–175, August 1999.

[12] T. Hara and M. Tomizuka, "Performance Enhancement of Multi-Rate Controller For Hard Disk Drives," *IEEE Transactions on Magnetics*, Vol. 35, No. 2, pp. 898–903, 1999.

[13] H. Fujimoto, Y. Hori, T. Yamaguchi, and S. Nakagawa, "Proposal of Perfect Tracking and Perfect Disturbance Rejection Control by Multirate Sampling and Applications to Hard Disk Drive Control," in *Proceedings of the IEEE CDC*, pp. 5277–5282, 1999.

[14] Y. Gu and M. Tomizuka, "Digital Redesign and Multirate Control For Motion Control—A General Approach and Application to Hard Disk Drive Servo System," in *Proceedings of the IEEE Internal Workshop Advanced Motion Control*, pp. 246–251, March 2000.

[15] K. Ohno and R. Horowitz, "A Multi-Rate Nonlinear State Estimator for Hard Disk Drives," in *Proceedings of the American Control Conference*, pp. 3083–3088, June 2003.

[16] M. Hirata, M. Takiguchi, and K. Nonami, "Track-Following Control of Hard Disk Drives Using Multi-Rate Sampled-Data H_∞ Control," in *Proceedings of the IEEE CDC*, pp. 3414–3419, 2003.

[17] S-. C. Wu and M. Tomizuka, "Performance and Aliasing Analysis of Multi-Rate Digital Controllers with Interlacing," in *Proceedings of the American Control Conference*, pp. 3514–3519, June 2004.

[18] L. Yang and M. Tomizuka, "Multi-Rate Short-Seeking Control of Dual-Actuator Hard Disk Drives for Computation Saving," in *Proceedings of the American Control Conference*, pp. 3210–3215, June 2005.

[19] T. Hara, "Open Loop Gain Criteria of Sampled-Data Control Systems With Mechanical Resonant Modes above the Nyquist Frequency," in *Proceedings of the SICE-ICASE International Joint Conference*, pp. 2771–2776, 2006.

[20] H. Fujimoto, F. Kawakami, and S. Kondo, "Multirate Repetitive Control and Applications—Verification of Switching Scheme by HDD and Visual Servoing," in *Proceedings of the American Control Conference*, pp. 2875–2880, 2003.

[21] M. Tomizuka, "Zero Phase Error Tracking Algorithm for Digital Control," *Transactions of ASME, Journal of Dynamic Systems, Measurement, and Control*, Vol. 109, pp. 65–68, March 1987.

[22] K. J. Åström, P. Hangander, and J. Sternby, "Zeros of Sampled System," *Automatica*, Vol. 20, No. 1, pp. 31–38, 1984.

[23] H. Fujimoto and A. Kawamura, "Perfect Tracking Digital Motion Control Based on Two-Degrees-of-Freedom Multirate Feedforward Control," in *Proceedings of IEEE International Workshop on Advanced Motion Control*, pp. 322–327, June 1998.

[24] H. Fujimoto, Y. Hori, and A. Kawamura, "Perfect Tracking Control Based on Multirate Feedforward Control With Generalized Sampling Periods," *IEEE Transactions on Industrial Electronics*, Vol. 48, No. 3, pp. 636–644, 2001.

[25] S. Nakagawa, T. Yamaguchi, H. Fujimoto, Y. Hori, K. Ito, and Y. Hata, "Multi-Rate Two-Degree-of-Freedom Control for Fast and Vibration-Less Seeking of Hard Disk Drives," in *Proceedings of the American Control Conference*, pp. 2797–2802, June 2001.

[26] H. Fujimoto, K. Fukushima, and S. Nakagawa, "Vibration Suppression Short-Span Seeking of HDD with Multirate Feedforward Control," in *Proceedings of the American Control Conference*, pp. 582–587, 2006.

[27] H. Fujimoto, "Short-Span Seeking Control of Hard Disk Drives with Multirate Vibration Suppression PTC," *Transactions of IEEJ*, Vol. 4, No. 2, pp. 184–191, 2009 (in Japanese).

[28] M. Hirata, K-. Z. Liu, and T. Mita, "Head Positioning Control of a Hard Disk Drive Using H_∞ Theory," in *Proceedings of the IEEE Control Decision Conference*, pp. 2460–2461, 1992.

[29] T. Atsumi, "Disturbance Suppression Beyond Nyquist Frequency in Hard Disk Drives," in *Proceedings of the IFAC World Congress*, pp. 827–832, 2008.

Chapter 5

Ultra-Precise Position Control

Takenori Atsumi

Hitachi, Ltd.

Mituso Hirata

Utsunomiya University

Hiroshi Fujimoto

The University of Tokyo

Nobutaka Bando

Japan Aerospace Exploration Agency

5.1   Phase-Stable Design for High Servo Bandwidth

The resonant characteristics of mechanical systems deteriorate the required achievable positioning accuracy in nanopositioning control systems. For instance, the lowest order mechanical resonant mode—hereinafter referred to as the *primary resonant mode*, limits the bandwidth of the control system while the higher order resonant modes—hereinafter referred to as the *high-order resonant modes*, affect stability in the high frequency range. The limitation on the bandwidth and stability affects positioning accuracy of the control systems. This directly implies that the performance of the control system is dependent on *how* these mechanical resonant modes are stabilized.

One of the methods used for stabilizing the mechanical resonant modes in controlled objects is the *gain-stable* design, where the open-loop gain is decreased to less than unity (0 dB) at frequencies of the mechanical resonant modes. In this method, notch filters are the most popular way to decrease the gain so that the control input does not excite the mechanical resonant modes. In fact, robust stability using H_∞ control theory and the small-gain theorem can also be considered a special case of the gain-stable design method.

On the other hand, the *phase-stable* design is widely used for head-positioning control systems in Hard Disk Drives (HDDs). In the phase-stable design, the phase of the open-loop characteristics is designed using the Nyquist diagram. In general, the designed controller based on the phase-stable design yields a better performance than that based on the gain-stable design.

In this section, a comparison between the gain-stable design and phase-stable design on the head-positioning control system in an HDD is presented.

5.1.1 Modeling of Controlled Object

The frequency response of the mechanical characteristics of the head-positioning system in an HDD is shown by the solid line in Figure 5.1. For such mechanical systems, a modeling method based on *modal superposition* is appropriate since the mechanical system consists of multiple resonant modes. In this method, the mechanical system of the controlled object $P_c(s)$ is assumed to be given by

$$P_c(s) = K_p \sum_{i=1}^{N} \frac{\kappa_i}{s^2 + 2\zeta_i \omega_{ni} s + \omega_{ni}^2}, \tag{5.1}$$

where K_p is the gain of the plant, N is the number of resonant modes considered, κ_i is the residue of each resonant mode, and ω_{ni} and ζ_i are the natural frequencies in rad/s and damping ratios of the resonant modes, respectively. These parameters are adjusted to ensure that the frequency response of $P_c(s)$ shown by the dashed line in Figure 5.1 coincides with the measured frequency response shown by the solid line in Figure 5.1. In this model, N is set as 5 (resulting in a tenth-order model), K_p is set at 3.7×10^7, and the values of the other parameters are listed in Table 5.1 [1]. Table 5.1 and Figure 5.1 indicate that large gains in the mechanical system are due to modes 2, 3, 4, while the gain contribution from mode 5 is relatively small. As such, the effects of mode 5 can be ignored.

TABLE 5.1: Parameters of $P_c(s)$.

mode i	mode 1	mode 2	mode 3	mode 4	mode 5
ω_{ni} (rad/s)	0	$2\pi \times 3950$	$2\pi \times 5400$	$2\pi \times 6100$	$2\pi \times 7100$
κ_i	1.0	−1.0	0.4	−1.2	0.9
ζ_i	0	0.035	0.015	0.015	0.060

The controlled object of the head-positioning control system in HDDs contains time delay elements due to the Zero-Order Hold (ZOH) and computation, etc., in addition to the characteristics of the head-positioning mechanical system shown in Figure 5.1. As such, the transfer function of the controlled object $P_d[z]$ consists of P_c discretized using the ZOH and the corresponding time delay [2]. In this case, the sampling time T_s and equivalent delay time T_d are set as

$$T_s = 50 \times 10^{-6} \text{ s}, \quad T_d = 15 \times 10^{-6} \text{ s}. \tag{5.2}$$

The frequency response of the controlled object is shown in Figure 5.2.

5.1.2 Controller Design Based on Vector Locus

In this section, an introduction to the relationship between the vector locus and the sensitivity transfer function is presented. The preliminary step

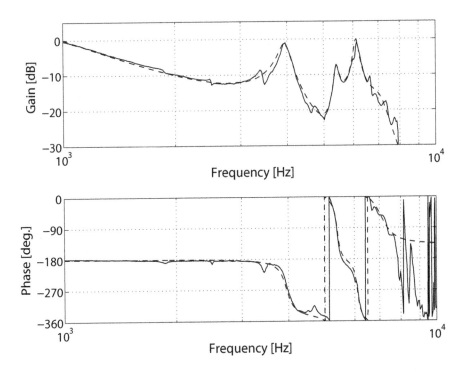

FIGURE 5.1: Frequency response of mechanical characteristics of the head-positioning system in HDDs.

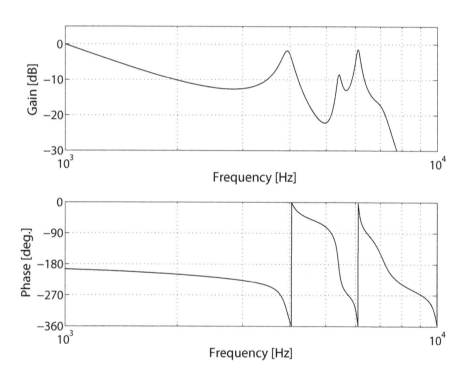

FIGURE 5.2: Frequency response of controlled object $P_d[z]$.

to designing a controller based on the analysis of the vector locus of the controlled object is detailed as well.

5.1.2.1 Relationship Between Vector Locus and Sensitivity Transfer Function

The *sensitivity transfer function* $S(j\omega)$ represents the characteristics of disturbance suppression on position and is the most important control performance index in control systems. In this section, we aim to design a control system which focuses on each resonant mode using the *vector locus* of the open-loop transfer function in a simple way.

We assume that the sensitivity transfer function $S(j\omega)$ is stable. The square of the gain of $S(j\omega)$ is given by

$$|S(j\omega)|^2 = \frac{1}{|1 + L(j\omega)|^2}, \tag{5.3}$$

where $L(j\omega)$ is the open-loop transfer function and ω is the angular frequency in rad/s. (5.3) can be rewritten as

$$|S(j\omega)|^2 = \frac{1}{(1 + a)^2 + b^2}, \tag{5.4}$$

where $a = \Re[L(j\omega)]$ and $b = \Im[L(j\omega)]$.

With $|R(j\omega)|$ defined as the distance from $L(j\omega)$ to the critical point $(-1, 0)$ on the complex plane, $|S(j\omega)|$ can also be expressed as

$$|S(j\omega)| = \frac{1}{|R(j\omega)|}. \tag{5.5}$$

Figure 5.3 shows the three-dimensional plot of $|S(j\omega)|$ where the x-y plane is the complex plane and the z-axis is the gain of $|S(j\omega)|$. It follows that the condition for suppressing positional disturbances is given by

$$|R(j\omega)| > 1, \tag{5.6}$$

and the condition that amplifies position disturbances is given by

$$|R(j\omega)| < 1. \tag{5.7}$$

Note that if $\angle L(j\omega)$ falls within $360n \pm 90°$ where n is an integer, the vector locus goes through the right-half plane of the Nyquist diagram and $|S(j\omega)| \leq 0$ is ensured regardless of the magnitude of the gain.

5.1.2.2 Vector Locus of Controlled Object

The vector locus of the mathematical model for the mechanical system is shown by the solid line in Figure 5.4. In Figure 5.4, the vector locus of the rigid body mode is the line from $-\infty$ towards the origin. The vector locus of

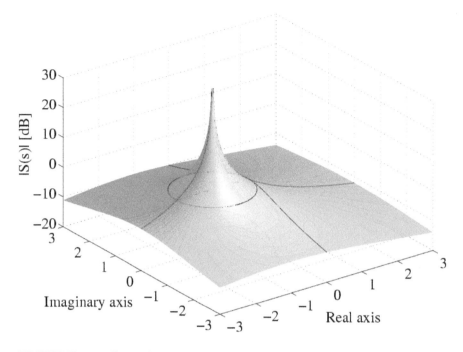

FIGURE 5.3: Gain of sensitivity transfer function in the Nyquist diagram.

mode 3 (which is a mechanical resonant mode with a positive residue) is the circle in the downward direction. The vector loci of the two resonant modes with negative residues (corresponding to modes 2 and 4) are the circles in the upward direction.

The vector locus of the controlled object (including the mechanical resonant modes, ZOH, and phase lag from the equivalent time delay) is shown by the dashed line in Figure 5.4 [2]. Figure 5.4 shows that the mechanical resonant modes with negative residues are in a stable condition as the vector loci recede from the critical point $(-1, 0)$ on the Nyquist diagram due to the phase delay of the control system. On the other hand, the mechanical resonant mode with the positive residue is in an unstable condition since the vector locus is near the critical point $(-1, 0)$ on the Nyquist diagram.

As such, the designed feedback controller should possess

- integral characteristics in the low frequency range for good steady-state and tracking performances;

- phase-lead characteristics in the open-loop transfer function around the gain crossover frequency to stabilize the rigid-body mode (mode 1);

- gain attenuation using notch filters for the resonant modes with positive residues; and

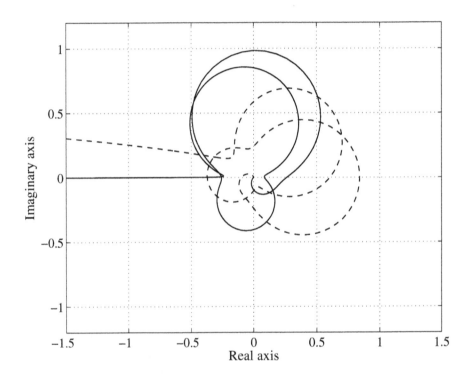

FIGURE 5.4: Vector loci of mechanical system and controlled object.

- phase stabilization of the mechanical resonant modes with negative residues (modes 2 and 4) without changing the phase characteristics around these resonant frequencies.

5.1.3 Controller Design

The feedback controller C is given by the product of the Proportional-Integral (PI) lead-lag filter C_s and the notch filter C_n. C_s provides the integral action and phase stabilizes the rigid-body mode. C_n stabilizes the mechanical resonant modes by decreasing the gain at the natural frequencies of the resonant modes. In this case, C_s is discretized using bilinear transformation and C_n is discretized using bilinear transformation with frequency prewarping [2]. As such, the feedback controller $C[z]$ is given as

$$C[z] = C_s[z]C_n[z], \tag{5.8}$$

and the open-loop transfer function $L[z]$ is given by

$$L[z] = P_d[z]C[z]. \tag{5.9}$$

5.1.3.1 Case 1: Gain-Stable Design for All Mechanical Resonant Modes

In this section, we consider the situation where all the resonant modes are gain stabilized by decreasing the gain at the natural frequencies with notch filters. Figure 5.2 shows that the controlled object has large gains around 3950 Hz, 5400 Hz, and 6100 Hz, corresponding to modes 2, 3, and 4, respectively. As such, the notch filter $C_n(s)$ is given by the product of three notch filters C_{n1}, C_{n2}, and C_{n3}, placed at 3950 Hz, 5400 Hz, and 6100 Hz, respectively, as

$$C_n(z) = C_{n1}(s)C_{n2}(s)C_{n3}(s), \tag{5.10}$$

where

$$C_{n1}(s) = \frac{s^2 + 2(0.02)(2\pi \times 3950)s + (2\pi \times 3950)^2}{s^2 + 2(0.3)(2\pi \times 3950)s + (2\pi \times 3950)^2}, \tag{5.11}$$

$$C_{n2}(s) = \frac{s^2 + 2(0.02)(2\pi \times 5400)s + (2\pi \times 5400)^2}{s^2 + 2(0.1)(2\pi \times 5400)s + (2\pi \times 5400)^2}, \tag{5.12}$$

$$C_{n3}(s) = \frac{s^2 + 2(0.02)(2\pi \times 6100)s + (2\pi \times 6100)^2}{s^2 + 2(0.3)(2\pi \times 6100)s + (2\pi \times 6100)^2}. \tag{5.13}$$

To increase the phase margin to more than 30° and the gain margin to more than 4.5 dB while ensuring that the open-loop gain above 3.8 kHz (resonant frequency of mode 2) is less than -10 dB, the PI lead-lag filter C_s is designed as

$$C_s(s) = \frac{3.19(s + 100 \times 2\pi)(s + 200 \times 2\pi)(s + 7000 \times 2\pi)}{s(s + 4400 \times 2\pi)(s + 5000 \times 2\pi)}, \quad (5.14)$$

so that the servo bandwidth is maximized. As a result, the servo bandwidth or the open-loop 0 dB gain cross frequency is 1000 Hz, with a phase margin of 30.2° and gain margin of 4.88 dB. The frequency response of the notch filter $C_n[z]$ is shown by the solid line and the frequency response of the PI lead-lag filter $C_s[z]$ is shown by the dashed line in Figure 5.5. The frequency response of the open-loop transfer function $L[z]$ is shown in Figure 5.6(a) and the corresponding Nyquist diagram is shown in Figure 5.6(b).

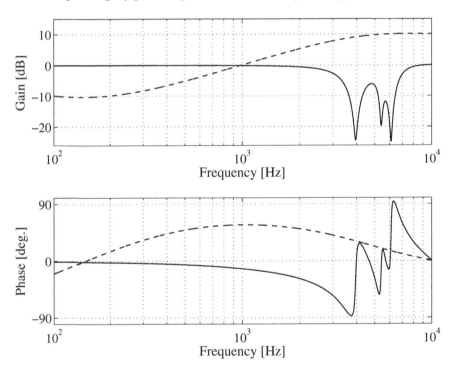

FIGURE 5.5: Frequency responses of controllers in case 1.

5.1.3.2 Case 2: Phase-Stable Design for Primary Mechanical Resonant Mode

In this section, we consider the situation where the primary resonant mode (mode 2) is phase stabilized while the other resonant modes (modes 3 and 4) are gain stabilized by notch filters. As such, the notch filter $C_n(s)$ is given by

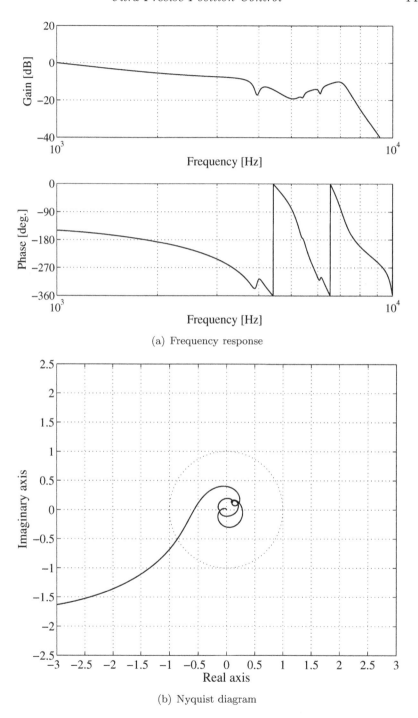

(a) Frequency response

(b) Nyquist diagram

FIGURE 5.6: Open-loop transfer function $L[z]$ in case 1.

148

the product of two notch filters C_{n2} and C_{n3} as

$$C_n(s) = C_{n2}(s)C_{n3}(s), \qquad (5.15)$$

where $C_{n2}(s)$ is placed at 5400 kHz as depicted in (5.12) and $C_{n3}(s)$ is placed at 6100 Hz as depicted in (5.13).

To increase the phase margin to more than 30° and the gain margin to more than 4.5 dB where ensuring that the open-loop gain above 5.4 kHz (resonant frequency of mode 3) is less than -10 dB, the PI lead-lag filter C_s is designed as

$$C_s(s) = \frac{2.54(s + 100 \times 2\pi)(s + 200 \times 2\pi)(s + 7000 \times 2\pi)}{s(s + 3500 \times 2\pi)(s + 4500 \times 2\pi)}, \qquad (5.16)$$

so that the servo bandwidth is maximized. As a result, the servo bandwidth is 1100 Hz, with the phase margin of 30.9° and the gain margin of 4.87 dB. The frequency response of the notch filter $C_n[z]$ is shown by the solid line and the frequency response of the PI lead-lag filter $C_s[z]$ is shown by the dashed line in Figure 5.7. The frequency response of the open-loop transfer function $L[z]$ is shown in Figure 5.8(a) and the corresponding Nyquist diagram is shown in Figure 5.8(b).

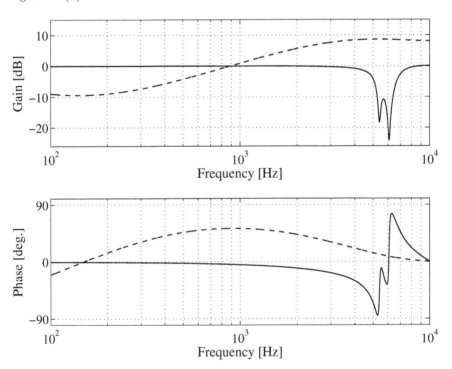

FIGURE 5.7: Frequency responses of controllers in case 2.

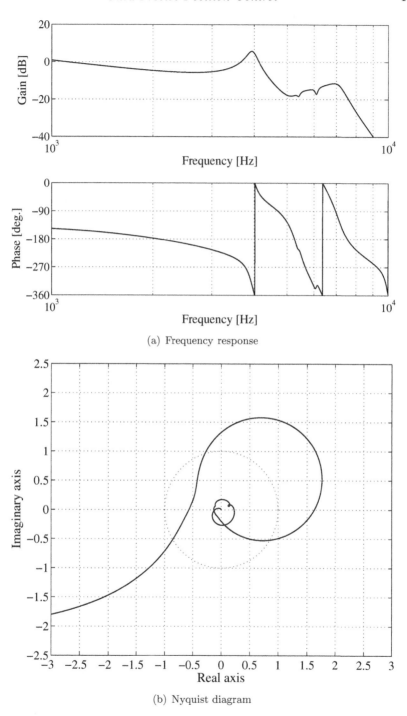

(a) Frequency response

(b) Nyquist diagram

FIGURE 5.8: Open-loop transfer function $L[z]$ in case 2.

5.1.3.3 Case 3: Phase-Stable Design for All Mechanical Resonant Modes

In this section, we consider the situation where the mechanical resonant modes with negative residues (modes 2 and 4) are phase stabilized while the other resonant mode (mode 3) is gain stabilized by a notch filter. As such, the notch filter $C_n(s)$ is identical to C_{n2} as

$$C_n(s) = C_{n2}(s), \qquad (5.17)$$

where $C_{n2}(s)$ is as depicted in (5.12).

To increase the phase margin to more than $30°$ and the gain margin to more than 4.5 dB while ensuring that the open-loop gain above 5.4 kHz (resonant frequency of mode 3) is less than -10 dB, the PI lead-lag filter C_s is designed as

$$C_s(s) = \frac{2.15(s + 100 \times 2\pi)(s + 200 \times 2\pi)(s + 7000 \times 2\pi)}{s(s + 3550 \times 2\pi)^2}, \qquad (5.18)$$

so that the servo bandwidth is maximized. As a result, the servo bandwidth is 1150 Hz, with the phase margin of $30.4°$ and the gain margin of 4.94 dB. The frequency response of the notch filter $C_n[z]$ is shown by the solid line and the frequency response of the PI lead-lag filter $C_s[z]$ is shown by the dashed line in Figure 5.9. The frequency response of the open-loop transfer function $L[z]$ is shown in Figure 5.10(a) and the corresponding Nyquist diagram is shown in Figure 5.10(b).

5.1.3.4 Comparison of Control Performances

In this section, we compare the control system for case 1, which is a gain-stable design for all mechanical resonant modes, case 2, which is a phase-stable design for the primary mechanical resonant mode only, and case 3, which is a phase-stable design for all mechanical resonant modes. The gains of the frequency responses of all three sensitivity transfer functions are shown in Figure 5.11, and the gains of the frequency responses of all three complementary sensitivity (or co-sensitivity) transfer functions are shown in Figure 5.12. In these figures, the results in cases 1, 2, and 3 are indicated by the dashed-dot, dashed, and solid lines, respectively.

The results of servo bandwidths, gain margins, phase margins, H_∞ norms of the sensitivity transfer functions, and H_∞ norms of the co-sensitivity transfer functions are shown in Table 5.2. These results show that the phase-stable design can improve the servo bandwidth by about 15% without any negative impacts on the stability margins of the control system. Also, the controllers can phase stabilize the mechanical resonant modes and improve the control performance while avoiding effects of phase delay below the center frequency when using conventional notch filters.

Additionally, the control system with phase-stable design has the following advantages:

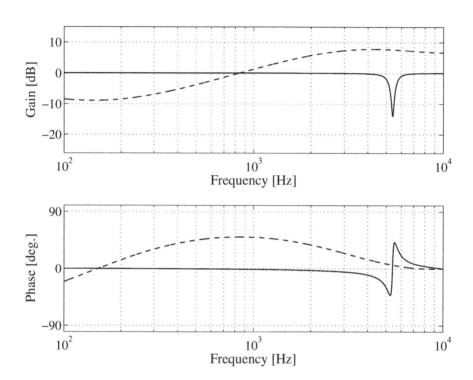

FIGURE 5.9: Frequency responses of controllers in case 3.

TABLE 5.2: Comparison of Control Performances

	Case 1	Case 2	Case 3
Servo bandwidth (Hz)	1000	1100	1150
Gain margin (dB)	4.88	4.87	4.94
Phase margin (deg)	30.2	30.9	30.4
$\|S\|_\infty$ (dB)	8.51	8.40	8.43
$\|T\|_\infty$ (dB)	6.02	5.87	6.02

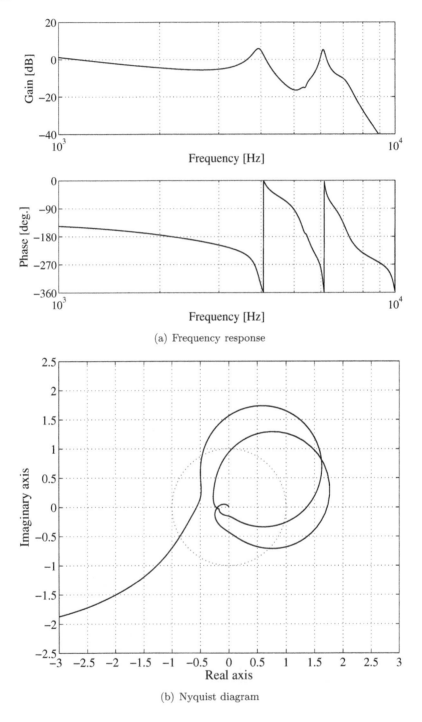

(a) Frequency response

(b) Nyquist diagram

FIGURE 5.10: Open-loop transfer function $L[z]$ in case 3.

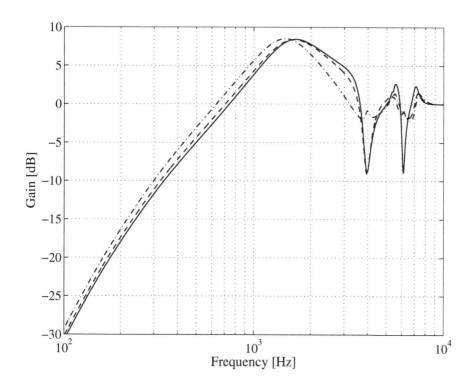

FIGURE 5.11: Gains of frequency responses of the sensitivity transfer functions.

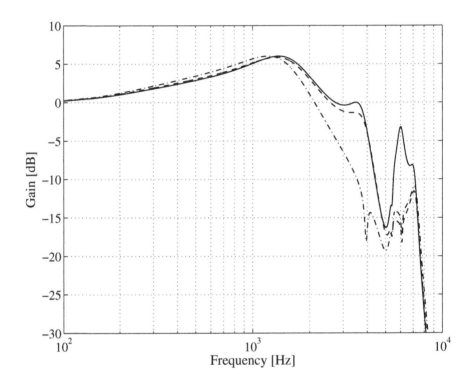

FIGURE 5.12: Gains of frequency responses of the complementary sensitivity transfer functions.

- **High robustness.** The phase-stable design is robust against variations in frequencies and gains since it can guarantee stability at the resonant frequencies as long as their vector loci are on the right-half plane of the Nyquist diagram. In contrast, the gain-stable design cannot guarantee stability when large variations in resonant frequencies occur; and

- **Reducing vibrations caused by mechanical resonances.** Vibrations caused by the mechanical resonant modes arise from flow induced vibrations, seek control, etc. The gain-stable design cannot suppress these vibrations since the notch filters remove these signals from the vibrations, but the phase-stable design can suppress them effectively in contrast.

5.2 Robust Control Using H_∞ Control Theory

In this section, the fundamentals of robust control using H_∞ control theory are introduced. The characterizations of the plant uncertainties, theory behind H_∞ control, as well as the formulation of various types of H_∞ control problems are presented. The effectiveness of using H_∞ control in HDDs is also verified.

5.2.1 Introduction

The control performance for mass-produced consumer mechatronics such as HDDs has to be improved as parametric variations of the mechanical actuators are inevitable. As such, the desired control performance should be achieved even when the properties of the plant are varied due to environmental or secular changes. This requirement for robustness and performance is generally conflicting, and is a fundamental issue which must be confronted when designing feedback control systems. In classical control theory, phase and gain margins are commonly used to evaluate the robustness of Single-Input-Single-Output (SISO) control systems to plant uncertainties. However, these metrics are insufficient as it is possible for the closed-loop system to be unstable, even when sufficient amounts of phase and gain margins exist.

In recent decades, advances in linear control theory provide a systematic methodology to design feedback control systems based on a mathematical model of the plant. An *optimal* controller is designed when a specific performance index is minimized during synthesis, and such a controller achieves good control performance when there are no plant uncertainties. As such, optimality is not guaranteed when variations in parameters or unmodeled dynamics are present. This lack of robustness in optimal control theories motivated much research interest in *robust* control theory, which provided a systematic way to cope with these plant uncertainties [3][4].

By far, the H_∞ control method is the most popular and well-known approach to designing a robust controller. In the H_∞ control method, a *robust stability condition* can be formulated using the H_∞ norm, and a robust controller that satisfies the robust stability condition can be obtained by solving two Algebraic Riccati Equations (AREs). In this method, specifications are defined in a frequency domain similar to transfer functions used in classical control theory. In addition, this theory also uses many linear systems techniques in the state-space domain [5][6]. Although the H_∞ control theory may be mathematically involved, the H_∞ controller design can be easily synthesized using Computer-Aided Control System Design (CACSD) tools such as MATLAB/Simulink® [7].

In this section, the plant is assumed to be a SISO system for simplicity but without loss of generality.

5.2.2   Mathematical Representation of Plant Uncertainties

In general, uncertainties in the plant will always exist and no mathematical model can represent a physical system exactly. Some sources of plant uncertainties and variations may include, but are not limited to

- **Modeling error**
  During mathematical modeling of a physical system, assumptions such as linearization, lumping of parameters, etc., are made. These discrepancies with the physical system cause modeling errors to be present.

- **Secular change**

- **Environmental change**
  Resistance changes due to variations in ambient temperature, changes in load, etc.

As such, it is common to model the physical systems plant as a *set* of plants instead. In this section, we describe the mathematical representation of plant uncertainties. The plant model without uncertainties is referred to as the *nominal model*.

5.2.2.1   Multiplicative Uncertainty

Let the transfer function of the nominal plant be P and that of the perturbed plant be $\tilde{P}$. An uncertainty Δ_m satisfying the following equation is referred to as *multiplicative uncertainty*

$$\tilde{P} = (1 + \Delta_m)P. \tag{5.19}$$

When $P(j\omega)$ and the frequency response $\tilde{P}(j\omega)$ of the actual system are given, Δ_m can be estimated as

$$\Delta_m(j\omega) = \frac{\tilde{P}(j\omega) - P(j\omega)}{P(j\omega)}. \tag{5.20}$$

5.2.2.2 Additive Uncertainty

Similarly, an uncertainty Δ_a satisfying the following equation is referred to as *additive uncertainty*

$$\tilde{P} = P + \Delta_a. \tag{5.21}$$

When $P(j\omega)$ and the frequency response $\tilde{P}(j\omega)$ of the actual system are given, Δ_a can be estimated as

$$\Delta_a(j\omega) = \tilde{P}(j\omega) - P(j\omega). \tag{5.22}$$

5.2.3 Robust Stability Problem

The controller design problem to obtain a controller that robustly stabilizes the closed-loop system for *all* uncertainties is known as the *robust stability problem*. We shall first introduce a well-known theorem which plays an important role in solving the robust stability problem.

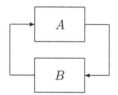

FIGURE 5.13: Small gain theorem.

Theorem 1 *Small Gain Theorem: Let $A(s)$ and $B(s)$ be stable and proper transfer functions shown in Figure 5.13. If*

$$|A(j\omega)B(j\omega)| < 1, \quad \forall\omega, \tag{5.23}$$

the closed-loop system in Figure 5.13 is stable.

In the small gain theorem, A and B can be unknown transfer functions, but their gains must be known. As such, it can be used to derive a robust stability condition for the closed-loop system with plant uncertainties.

We will derive a robust stability condition for the closed-loop system shown in Figure 5.14(a) when the plant has a multiplicative uncertainty. The multiplicative uncertainty is assumed to be stable and proper for simplicity but without loss of generality.

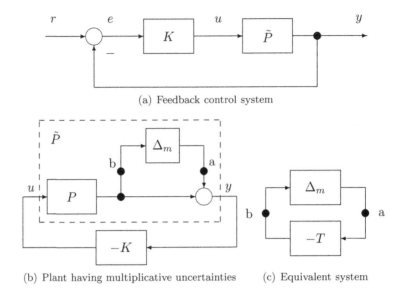

(a) Feedback control system

(b) Plant having multiplicative uncertainties   (c) Equivalent system

FIGURE 5.14: Robust stabilization for multiplicative uncertainties.

It can be easily verified that the closed-loop system in Figure 5.14(a) can be transformed to an equivalent system shown in Figure 5.14(b) by omitting r, which is immaterial in stability analysis. As such, the transfer function from "a" to "b" is $-T$, where

$$T := \frac{PK}{1 + PK}. \tag{5.24}$$

Now, Figure 5.14(b) can be transformed to the equivalent system shown in Figure 5.14(c). By applying the small gain theorem to Figure 5.14(c), we have a robust stability condition for the plant with multiplicative uncertainties as

$$|\Delta_m(j\omega)T(j\omega)| < 1, \quad \forall \omega. \tag{5.25}$$

Since (5.25) has an unknown transfer function Δ_m, it cannot be used for controller design. As such, we introduce a known stable transfer function $w_m(s)$ that satisfies

$$|\Delta_m(j\omega)| \le |w_m(j\omega)|, \quad \forall \omega. \tag{5.26}$$

The *sufficient* condition for robust stability of the plant with multiplicative uncertainties can now be obtained as

$$|w_m(j\omega)T(j\omega)| < 1, \quad \forall \omega. \tag{5.27}$$

It is obvious that (5.25) is satisfied if (5.26) holds. The transfer function T in (5.27) is also commonly referred to as the *complementary sensitivity transfer function*.

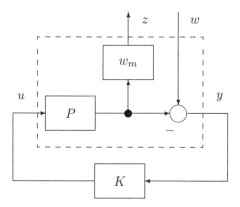

FIGURE 5.15: Robust stabilization for multiplicative uncertainty.

5.2.4 H_∞ Control Theory

As mentioned in previous sections, the robust stabilization problem for the plant with multiplicative uncertainties is formulated as a controller design problem which satisfies (5.27). Such a controller can be calculated by using H_∞ *control theory*.

For a stable and proper transfer function $G(s)$, the H_∞ *norm* is defined as

$$\|G(s)\|_\infty := \sup_{0 \le \omega \le \infty} \bar{\sigma}\{G(j\omega)\}, \qquad (5.28)$$

where $\bar{\sigma}$ is the maximum singular value of a complex matrix A and is defined as

$$\bar{\sigma}(A) = \sqrt{\lambda_{max}(A^*A)}. \qquad (5.29)$$

Note that A^* is the conjugate transpose of A, and λ_{max} is the maximum value of eigenvalues. If $G(s)$ is a SISO system, (5.28) takes on a simpler form as

$$\|G(s)\|_\infty := \max_{0 \le \omega \le \infty} |G(j\omega)|, \qquad (5.30)$$

i.e., the H_∞ norm corresponds to the *maximum gain* of $G(s)$. Furthermore, the following inequality

$$\|G(s)\|_\infty < 1 \Leftrightarrow |G(j\omega)| < 1, \quad \forall \omega \qquad (5.31)$$

obtained from (5.30) is commonly used when the gain condition is substituted for the H_∞ norm condition.

Using (5.31), the robust stability condition of (5.27) can be represented by the H_∞ norm condition of

$$\|w_m T\|_\infty < 1. \qquad (5.32)$$

(5.32) implies that the H_∞ norm from w to z in Figure 5.15 should be less

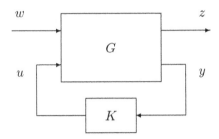

FIGURE 5.16: Generalized plant.

than one. In H_∞ control, various control problems are formulated using H_∞ norm conditions.

For generality, we introduce a generalized plant G, as shown in Figure 5.16. The box surrounded by the dashed line in Figure 5.15 corresponds to G in Figure 5.16, and G is referred to as the *generalized plant*. The input and output relations of $G(s)$ are described as

$$\left[\begin{array}{c} z \\ y \end{array}\right] = G \left[\begin{array}{c} w \\ u \end{array}\right], \tag{5.33}$$

where

$$G =: \left[\begin{array}{cc} G_{11} & G_{12} \\ G_{21} & G_{22} \end{array}\right]. \tag{5.34}$$

The vector signals w, z, u, and y are referred to as the *exogenous input*, *controlled output*, *control input*, and *measurement output*, respectively.

The H_∞ control problem is formulated as follows.

Definition 1 (H_∞ control problem) *Consider a closed-loop system with a generalized plant G and feedback controller K, as shown in Figure 5.16. The objective of H_∞ control is to find a controller K such that the closed-loop system is internally stable and the H_∞ norm condition*

$$\|G_{zw}\|_\infty < \gamma \tag{5.35}$$

is satisfied for a given positive $\gamma \in \mathcal{R}$, where G_{zw} is the closed-loop transfer function from w to z defined as

$$G_{zw} = G_{11} + G_{12}K(I - G_{22}K)^{-1}G_{21}. \tag{5.36}$$

Let us define a state-space representation of G as

$$\dot{x} = Ax + B_1 w + B_2 u, \tag{5.37}$$

$$z = C_1 x + D_{11} w + D_{12} u, \tag{5.38}$$

$$y = C_2 x + D_{21} w. \tag{5.39}$$

The H_∞ control problem under the following assumptions is called the *standard H_∞ control problem* [8], and the H_∞ controller can be readily calculated using MATLAB®.

Assumption 1 A1 (A, B_2) *is stabilizable and* (C_2, A) *is detectable.*

A2 D_{12} *is full column rank and* D_{21} *is full row rank.*

A3 G_{12} *and* G_{21} *have no invariant zeros on the* $j\omega$*-axis, i.e.,*

$$\left[\begin{array}{cc} A - j\omega I & B_2 \\ C_1 & D_{12} \end{array} \right] \tag{5.40}$$

has full column rank for all ω*, and*

$$\left[\begin{array}{cc} A - j\omega I & B_1 \\ C_2 & D_{21} \end{array} \right] \tag{5.41}$$

has full row rank for all ω*.*

The solution to the standard H_∞ control problem is obtained by solving two AREs. The solution can also be obtained using the Linear Matrix Inequality (LMI) as well. The LMI-based solution requires only Assumption A1 to hold [9].

5.2.5 Various H_∞ Control Problems

The H_∞ control theory can be used to meet various closed-loop performance specifications. In this section, the formulations of the H_∞ *sensitivity minimization* and H_∞ *mixed sensitivity* problems are presented.

5.2.5.1 Sensitivity Minimization Problem

The closed-loop transfer function from r to y in Figure 5.14(a) is also known as the *complementary sensitivity transfer function*, i.e., $y = Tr$, and determines the tracking performance of the control system to reference input r.

In a feedback control system, it is desired that the tracking performance be unaffected by variations in the plant. In order to quantify how sensitive T is to plant variations, we take the limiting ratios of a relative perturbation of T, i.e., $\Delta T/T$, to a relative perturbation of P, i.e., $\Delta P/P$, as

$$\lim_{\Delta P \to 0} \frac{\Delta T/T}{\Delta P/P} = \frac{dT}{dP} \frac{P}{T} = \frac{1}{1 + PK} =: S. \tag{5.42}$$

S is referred to as the *sensitivity transfer function*, and it corresponds to the transfer function from r to e. To achieve a smaller tracking error for any reference input r, the gain of S should be small at all frequencies. However, $P(\infty) = 0$ holds since P is strictly proper in general, which implies that

$S(\infty) = \frac{1}{1+P(\infty)K(\infty)} = 1$. As such, it is impossible to minimize the gain of S for all frequencies. However, the reference input r has low frequency components in general, and hence it is sufficient to minimize S only in the low frequency range. This requirement can be formulated as

$$|S(j\omega)| < 1/|w_S(j\omega)|, \quad \forall\omega, \qquad (5.43)$$

where w_S is a stable and proper transfer function. If (5.43) is satisfied, the gain of S lies under the gain plot of $1/w_S$. Thus, an ideal frequency response of S can be expressed in terms of the weighting function w_S, and (5.43) is described by the H_∞ norm condition as

$$|S(j\omega)| < 1/|w_S(j\omega)| \quad \Leftrightarrow \quad |w_S(j\omega)S(j\omega)| < 1$$
$$\Leftrightarrow \quad \|w_S S\|_\infty < 1. \qquad (5.44)$$

5.2.5.2 Mixed Sensitivity Problem

During controller design and synthesis, robust stability itself is insufficient in many realistic situations. For example, a passive system such as the mass-spring-damper is always stable regardless of how much the plant parameters vary, and an intuitive robust stabilizing controller candidate will be $K = 0$. As such, it is important to design feedback controllers to have *both* good performance and robust stability. One way to achieve this goal is to obtain a feedback controller such that *both* robust stability and sensitivity conditions are satisfied, i.e., (5.32) and (5.44) hold. This problem is known as the *mixed sensitivity problem*.

In H_∞ control problems, it is not possible to treat two norm conditions separately. As such, the following sufficient condition is generally used

$$\left\| \begin{array}{c} w_S S \\ w_m T \end{array} \right\|_\infty < 1. \qquad (5.45)$$

The generalized plant in (5.45) is shown in Figure 5.17, and the transfer functions from w to z_1 and z_2 are $w_S S$ and $w_m T$, respectively.

It is easy to show that the following identity holds for any S and T

$$S + T = 1. \qquad (5.46)$$

(5.46) implies that it is impossible to minimize the gain of S and T simultaneously at all frequencies. As such, S is minimized at low frequencies where the frequency spectrum of r is, and T is minimized at high frequencies where the frequency spectra of plant uncertainties are. It is worth noting that the name "complementary sensitivity" comes from (5.46).

5.2.6 Application of H_∞ Control to HDDs

In this section, an H_∞ controller for track-following control systems in HDDs is designed. The head actuator of an HDD can be modeled as a plant

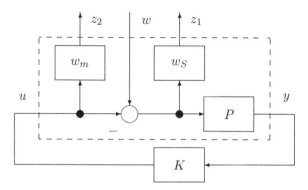

FIGURE 5.17: Mixed sensitivity problem.

which has a rigid body mode and mechanical resonant modes at high frequency. If the transfer function of the plant is exactly known and the property does not change at all, a high control performance can be obtained. However, it is difficult to obtain an exact model in practical situations, and the plant properties also vary among various products during mass production of HDDs.

Here we consider the design of an H_∞ controller such that the high frequency resonant modes are regarded as uncertainties and the closed-loop system is robustly stable to these uncertainties [10].

The frequency response of the plant that was obtained by a servo analyzer is shown by the solid line in Figure 5.18. The nominal model of the plant is defined as a rigid body mode system with no viscous friction as

$$P = \frac{k_p}{s^2}. \tag{5.47}$$

The multiplicative uncertainty Δ_m is calculated using (5.20) and is shown by the dashed line in Figure 5.19. The weighting function for robust stability is defined by a fourth-order transfer function as

$$w_m(s) = \left[\frac{s^2 + 2\zeta_n\omega_{n2}s + \omega_{n2}^2}{s^2 + 2\zeta_d\omega_{d2}s + \omega_{d2}^2} \right]^2 \times g_2. \tag{5.48}$$

The parameters of w_m are selected as $\omega_{d2} = 6400\pi$, $\zeta_d = 0.3$, $\omega_{n2} = 2400\pi$, $\zeta_n = 0.66$, and $g_2 = 29.1$ for the upper bound Δ_m, as shown by the solid line in Figure 5.19.

It is well known that the mixed sensitivity problem does not satisfy the assumptions of the standard H_∞ problem if the plant has poles on the $j\omega$ axis. Since (5.47) has two poles at the origin, a modification is required. The track-following control problem can be directly formulated into the H_∞ control problem if disturbance suppression is important. This can be achieved by minimizing the H_∞ norm of the closed-loop transfer function from d to y in

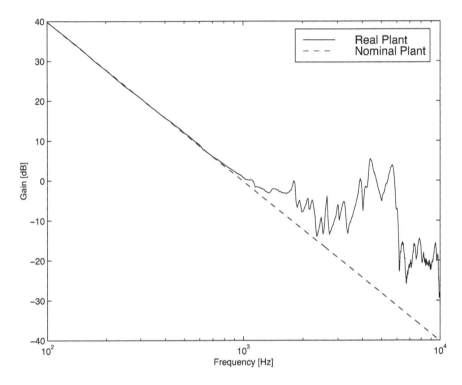

FIGURE 5.18: Frequency response of plant.

Figure 5.20. Since the transfer function from d to y is

$$y = \frac{P}{1 + PK}d = PSd, \tag{5.49}$$

the H_∞ norm condition is formulated as

$$\|w_S PS\|_\infty < 1, \tag{5.50}$$

where w_S is a weighting function to PS. Finally, the mixed sensitivity problem is modified as

$$\left\| \begin{array}{c} w_S PS \\ w_m T \end{array} \right\|_\infty < 1. \tag{5.51}$$

The generalized plant for (5.51) is shown in Figure 5.21(a). However, this generalized plant does not satisfy assumption **A2** of the standard H_∞ problem because the direct feedthrough from w to y is zero. Therefore, the generalized plant is slightly modified, as shown in Figure 5.21(b), by introducing another exogenous input w_2 and a small positive value ϵ. It is obvious that the H_∞ norm from w to z in Figure 5.21(a) is less than one if that in Figure 5.21(b) is also less than one.

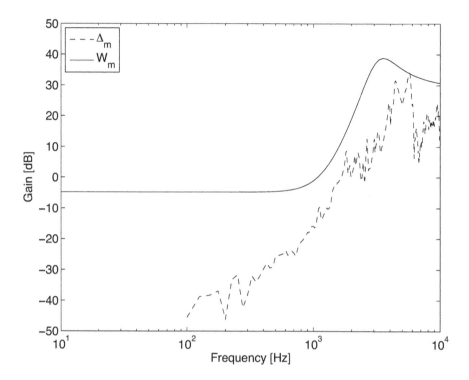

FIGURE 5.19: Frequency responses of Δ_m and w_m.

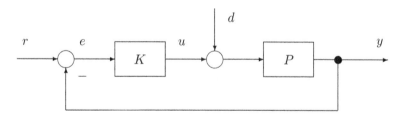

FIGURE 5.20: Track-following control system.

The weighting function for disturbance attenuation is selected to have a large gain at the low frequency region where the disturbance spectrum is large. The trade-off of $S + T = 1$ has to be also considered in weight selection. After some trial and error, we have

$$w_S(s) = \frac{s + \omega_{n1}}{s + \omega_{d1}} \times g_1, \tag{5.52}$$

where $\omega_{n1} = \pi$, $\omega_{d1} = 1 \times 10^{-6}$, and $\epsilon = 0.04$.

An H_∞ controller was calculated by using MATLAB®, and its frequency response is shown in Figure 5.22. The minimum H_∞ norm was 0.9821. The H_∞

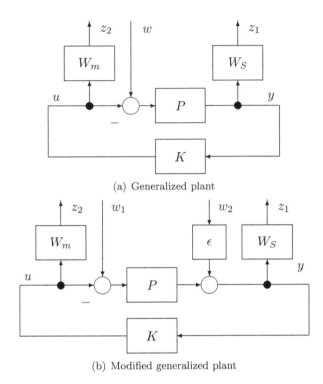

(a) Generalized plant

(b) Modified generalized plant

FIGURE 5.21: Generalized plants.

controller was discretized by Tustin transformation with a sampling period of 50 μs, and a step reference and impulse disturbance response were obtained, as shown in Figure 5.23. The impulse disturbance was injected at the plant input when $t = 5$ ms. The step response has a large overshoot but the disturbance response is satisfactory. Since the H_∞ controller design is performed in the frequency domain, it is difficult to incorporate time domain specifications such as reduction of overshoot. However, it should be noted that it is straightforward to employ Two-Degrees-of-Freedom (TDOF) control methods to improve the step response when using H_∞ control. It is recommended to design the feedback controller to have good disturbance attenuation and robust stability since the reference response can be easily improved using TDOF control methods.

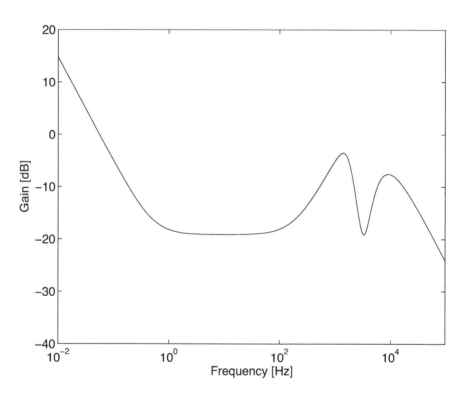

FIGURE 5.22: Magnitude response of H_∞ controller.

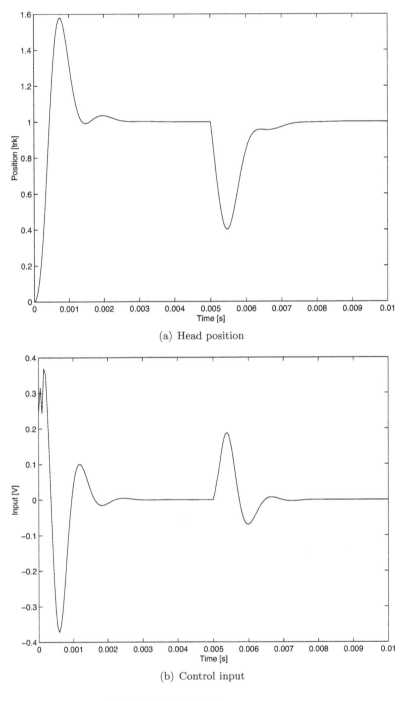

(a) Head position

(b) Control input

FIGURE 5.23: Time responses.

5.3 Multi-Rate H_∞ Control

In this section, the multi-rate discrete-time H_∞ control and sampled-data H_∞ control methods are introduced to design H_∞ controllers [11].

5.3.1 Multi-Rate Discrete-Time H_∞ Control

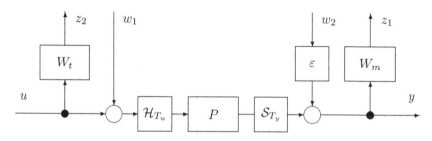

FIGURE 5.24: Generalized plant for multi-rate discrete-time H_∞ controller design.

In the discrete-time H_∞ control problem, the generalized plant is defined in the discrete-time domain and a discrete-time controller is designed. As such, it is unnecessary to discretize the controller before implementation in digital control systems. When it is not possible to use a high sampling rate (as compared to the control bandwidth), the discrete-time H_∞ method may have a better performance as compared to the continuous-time H_∞ method. However, all the weighting functions will have to be reselected in the discrete-time domain. In this section, a method to design multi-rate controllers using the discrete-time H_∞ control method is introduced.

To design a track-following controller, the generalized plant shown in Figure 5.24 is considered. In this figure, $\mathcal{H}_{T_u}$ is the hold and $\mathcal{S}_{T_y}$ is the sampler. W_t, W_m, and ε are the weights. The sampling periods of $\mathcal{H}_{T_u}$ and $\mathcal{S}_{T_y}$ are represented by T_u and T_y, respectively, and $T_y = 200$ μs is assumed.

Next, we design a single-rate controller with $T_u = 200$ μs and a multi-rate controller with $T_u = \frac{T_y}{2} = 100$ μs. In the generalized plant shown in Figure 5.24, the disturbance attenuation performance is defined by the signals in the path from w_1 to z_1, while the performance of robust stability against multiplicative perturbation is defined by the signals in the path from w_1 to z_2. It is worth noting that the sampling periods of W_t and W_m are different when a multi-rate controller is designed. W_m is designed to have a larger gain at the low frequency region for good disturbance attenuation, while W_t is designed to be an upper bound of the multiplicative perturbation for robust stability. These weighting functions are chosen using discrete-time transfer functions.

The frequency responses of the single-rate and multi-rate weighting functions W_t as well as the multiplicative perturbation Δ_m are shown in Figure 5.25.

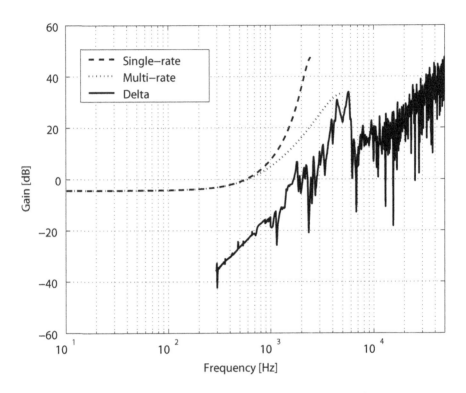

FIGURE 5.25: Frequency responses of W_t and Δ_m.

For the single-rate case, the H_∞ controller can be easily synthesized using the built-in function in the MATLAB® toolbox. On the other hand, the generalized plant with multiple sampling periods is not a Linear Time-Invariant (LTI) system and the multi-rate controller cannot be readily obtained in the multi-rate case. As such, the generalized plant is transformed into an LTI system using the discrete-time *lifting technique*. As the multi-rate ratio is two in this design, the LTI generalized plant with a lifted input signal $\underline{u}$ of size two is obtained, as shown in Figure 5.26.

It follows from the fact that the H_∞ norm is invariant under discrete-time lifting [12] and the LTI H_∞ controller which minimizes the H_∞ norm from w to z in Figure 5.26 is also a solution of the original problem.

The lifted plant $\underline{P}_d$ in Figure 5.26 is obtained from (3.103)–(3.107) with $M = 2$. The state-space representation of $\underline{W}_t$ is given as

$$
\underline{W}_t = \left[
\begin{array}{c|cc}
A_t^2 & A_t B_t & B_t \\
\hline
C_t & D_t & 0 \\
C_t A_t & C_t B_t & D_t
\end{array}
\right],
\tag{5.53}
$$

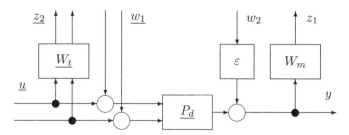

FIGURE 5.26: Lifted generalized plant.

where A_t, B_t, C_t, and D_t are the state-space matrices of W_t.

Track-following experiments were carried out by using the obtained H_∞ controllers. The results are shown in Figure 5.27. The 3σ value of the tracking error is improved from 0.0437 of a track in the single-rate case to 0.0417 of a track in the multi-rate case using the multi-rate H_∞ controller.

5.3.2 Multi-Rate Sampled-Data H_∞ Control

The two well-known approaches for designing a digital controller are

(1) **Continuous-time design and controller discretization**
A continuous-time controller is designed using continuous-time control theory, and a digital controller is obtained via discretizing the continuous-time controller using Tustin transformation, etc.; and

(2) **Discrete-time design**
The plant is first discretized and the digital controller is designed using discrete-time control theory.

Method (1) can achieve the same control performance as compared to that of the original continuous-time controller when the sampling rate is high enough as compared to the control bandwidth. However, when the sampling rate is not so fast, closed-loop stability may not be guaranteed. In Method (2), closed-loop stability is always ensured. However, the performance is only ensured at the sampling instants since the inter-sampling information is lost when the plant is discretized. In some cases, inter-sampling oscillations or *ripples* are observed.

In order to solve these problems, the sampled-data H_∞ control theory has been proposed since sampled-data control theory is well known. In the sampled-data H_∞ control theory, the generalized plant shown in Figure 5.28 is used. The main feature of the sampled-data H_∞ control theory is that the signals w, u, z, and y are defined in the continuous-time domain while the controller is defined in the discrete-time domain. Since the closed-loop system shown in Figure 5.28 is a *hybrid* system which consists of both continuous- and discrete-time signals, an L_2-induced norm [12] is used for the norm criterion

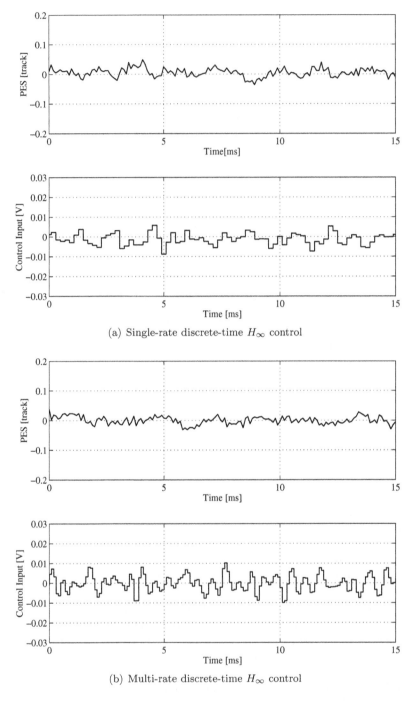

(a) Single-rate discrete-time H_∞ control

(b) Multi-rate discrete-time H_∞ control

FIGURE 5.27: Position Error Signals (PES) and control inputs.

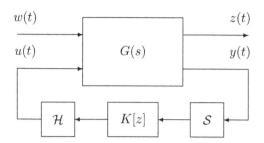

FIGURE 5.28: Single-rate and multi-rate sampled-data H_∞ control problem.

from w to z as

$$J = \sup_{w(t) \in \mathcal{L}^2} \frac{\|z(t)\|_2}{\|w(t)\|_2}. \tag{5.54}$$

It is worth noting that the L_2-induced norm and the H_∞ norm are equivalent when the system is LTI.

As such, the sampled-data H_∞ control problem can be reformulated as a problem to design a discrete-time controller $K[z]$ such that the discrete-time controller stabilizes the continuous-time generalized plant G and minimizes the L_2-induced norm from w to z. The most distinctive feature of this approach is that the discrete-time controller which minimizes the L_2-induced norm can be directly obtained by evaluating the controlled variable $z(t)$ over the sampling interval, while the disturbance $d(t)$ is injected not only at the sampling instants but within the sampling intervals as well. Traditionally, sampled-data H_∞ control theory is very mathematically involved and is difficult to obtain the solution. However, with the release of the Sampled-Data Control Toolbox in MATLAB® by leading researchers in this field, it is becoming a reality that sampled-data H_∞ controller design can be easily carried out [13].

When the sampling periods of the sampler $\mathcal{S}$ and the hold $\mathcal{H}$ in Figure 5.28 are different, the sampled-data H_∞ control problem can be reformulated as a *multi-rate H_∞ control problem*. Similar to the multi-rate discrete-time H_∞ control problem, the multi-rate sampled-data H_∞ control problem can be solved by transforming it to a single-rate sampled-data H_∞ control problem using the lifting technique [14].

Next, we design a multi-rate H_∞ controller for track-following control of an HDD. The generalized plant is shown in Figure 5.29. Robust stability to additive uncertainty is considered by the path from w_1 to z_1 with the weighting function W_t. It is noted that d and d^{-1} are scaling parameters introduced to relax the robust stability condition. The norm from w_2 to z_2 is evaluated with the weighting function W_s when considering the disturbance attenuation performance.

In multi-rate control, the control input is updated several times during the sampling interval of y. As such, the control input may have high frequency

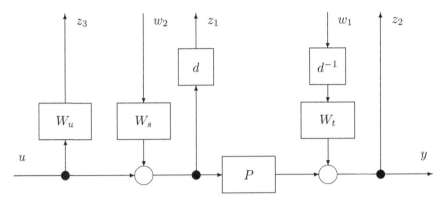

FIGURE 5.29: Generalized plant for multi-rate sampled-data H_∞ control problem.

components, and high frequency resonant modes of the plant might be excited if the control input is not taken into account. The smoothness of the control input is an important factor in multi-rate control design, especially for systems having mechanical resonant modes, as discussed in [15]. Therefore, an additional control variable z_3 is introduced to shape the frequency components of the control input.

The frequency responses of W_s, W_t, and W_u with additive uncertainties are shown in Figure 5.30. The weighting function W_t is chosen to bound the additive perturbation as tightly as possible since the robust stability condition is conservative in sampled-data systems. The orders of W_s, W_t, and W_u are 1, 7, and 2, respectively. The scaling parameter d is adjusted to minimize γ for fixed W_t, W_s, and W_u.

The frequency response of a multi-rate controller cannot be calculated in a standard manner as it is not an LTI system. Thus, a method based on a polyphase representation is introduced here. In order to obtain a frequency response up to the Nyquist frequency of $\frac{1}{2T_u}$, we consider the block diagram in Figure 5.31, where K_1 and K_2 are the elements of the lifted controller $\underline{K}$ given by

$$\underline{K} = \begin{bmatrix} K_1 \\ K_2 \end{bmatrix}. \tag{5.55}$$

The representation

$$K_{poly}[z] = K_1[z^2] + z^{-1}K_2[z^2] \tag{5.56}$$

is known as the *polyphase representation*. The multi-rate sampled-data H_∞ controller was calculated based on the method described in [14]. The frequency response of the obtained controller is shown by the solid line in Figure 5.32.

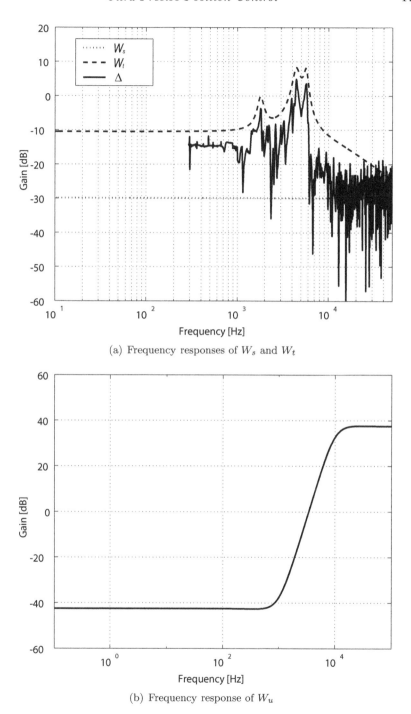

(a) Frequency responses of W_s and W_t

(b) Frequency response of W_u

FIGURE 5.30: Frequency responses of weighting functions.

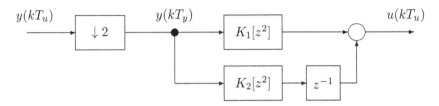

FIGURE 5.31: Polyphase representation of multi-rate controller.

Since the gain of an m-multiple down-sampler is $1/m$, the frequency response $\hat{K}(j\omega)$ in Figure 5.31 is calculated as

$$\hat{K}(j\omega) = \frac{1}{2} K_{poly}[e^{j\omega T_u}].$$
(5.57)

It is worth noting that the aliasing effect is not taken into account in the frequency response when calculating $\hat{K}(j\omega)$ using (5.57).

For comparison purposes, a single-rate sampled-data H_∞ controller was designed assuming $T_u = T_y = 200 \ \mu$s. In the single-rate design, W_u was not used, i.e., $W_u = 0$ was assumed while the same weighting functions and parameters of the multi-rate design were used. The obtained single-rate controller is shown by the dashed line in Figure 5.32. Figure 5.32 indicates that both the single-rate and multi-rate controllers are suppressing the gain around the frequency where aliasing of the high frequency resonant mode components exists. As such, it is verified that the aliasing effect is automatically taken into account in the sampled-data design.

Similarly, track-following experiments were carried out using the single-rate and multi-rate H_∞ controllers, and the corresponding PES and control inputs are shown in Figure 5.33. From these figures, it can be observed that the tracking errors and maximum amplitudes of the control inputs are reduced by the multi-rate controller. The 3σ value of 0.0413 of a track in the single-rate case is reduced to 0.0399 of a track in the multi-rate case. By comparing the power spectrum densities of the PES, it can also be seen that the tracking error is improved by the multi-rate control at the low frequency region.

Besides the multi-rate design example described here, many other multi-rate controller design approaches have been proposed in HDD control. The multi-rate sampled-data H_∞ control was applied to a dual-stage actuator in [16], and a single-rate H_∞ controller was implemented as a multi-rate controller by using a zero-order interpolator in [17].

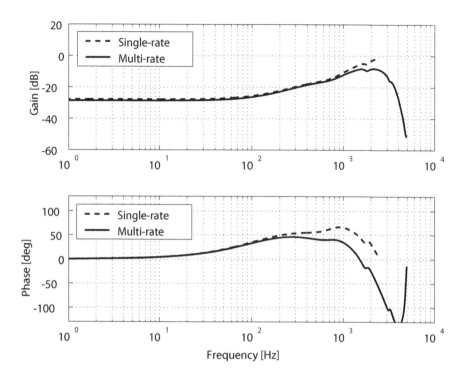

FIGURE 5.32: Frequency responses of multi-rate and single-rate sampled-data controllers.

5.4   Repetitive Control

In this section, the *Repetitive Perfect Tracking Control* (RPTC) methodology is proposed to reject high-order disturbance modes. RPTC is applied to track-following control in HDDs to improve their track-following performances, and its effectiveness is verified through simulations and experimental implementations on actual HDDs.

5.4.1   Introduction

In the head-positioning system of HDDs, the head position is detected by discrete servo signals embedded in the disks. As such, the sampling period of the PES is restricted by the number of samples and rotational frequency of the spindle motor.

Repetitive control is a widely used technique for rejecting periodic disturbances or tracking of periodic reference signals [18][19]. Although this control

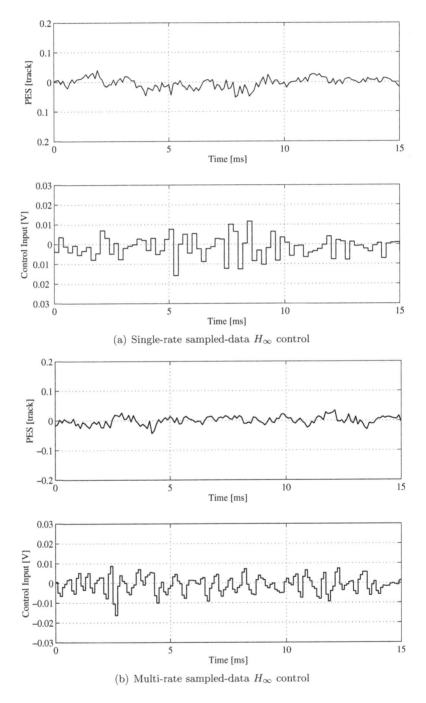

(a) Single-rate sampled-data H_∞ control

(b) Multi-rate sampled-data H_∞ control

FIGURE 5.33: PES and control inputs.

scheme has excellent performance for low order disturbance modes, it cannot reject relatively higher frequency modes. The reasons are:

1. the delay caused by the ZOH at the plant input is significant when the higher frequency modes to be rejected are close to the Nyquist frequency;

2. a Low Pass Filter (LPF) is required to maintain the stability robustness; and

3. an approximation of the inverse transfer function is required to deal with the unstable zero of the discrete-time plant in the conventional discrete-time repetitive controller [18].

However, these problems can be overcome by introducing a control scheme called *Repetitive Perfect Tracking Control* (RPTC) [20][21].

Recent developments in control processors and interface technologies allow shortening of the control period T_u to less than the sampling period T_y when the sampling rate of the sensor is restricted. The above-mentioned problem (1) can be solved by using the multi-rate input control where $T_u < T_y$.

In conventional digital repetitive control, the discrete-time disturbance model $(z^{N_d} - 1)^{-1}$ is implemented in a feedback loop using the internal model [18]. Although the sensitivity transfer function becomes zero at the frequencies of the disturbance harmonics, the sensitivity transfer function has large amplitude at other frequencies, which results in severe reduction of total tracking accuracy. Moreover, the closed-loop system may be unstable when the peak gain of the internal model excites the unmodeled dynamics. As such, an LPF is usually implemented in repetitive control to assure the robust stability with a tradeoff of high frequency disturbance rejection performance. On the other hand, novel switching schemes to achieve repetitive disturbance rejection by feedforward control are also introduced in [20][21].

The above-mentioned problem (3) of discrete-time unstable zero is not crucial in conventional feedback repetitive control as the stability can be ensured when using the approximate Zero-Phase Error (ZPE) inverse [18]. However, the gain characteristics of the ZPE result in tracking error, especially for high order disturbances when the feedforward scheme is introduced with a switching scheme [22]. This motivates the use of Perfect Tracking Control (PTC) with multi-rate inputs to obtain the ideal inner-loop system in the discrete-time domain [20][21].

5.4.2 Repetitive Perfect Tracking Control (RPTC)

In this section, the concepts of RPTC are introduced using a discrete-time plant model with multi-rate hold and PTC design.

5.4.2.1 Discrete-Time Plant Model with Multi-Rate Hold

In this section, it is assumed that the control input can be changed N times during the sampling period of output signal T_y. For simplicity, but without

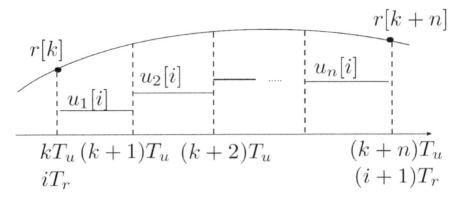

FIGURE 5.34: Multi-rate hold.

loss of generality, the input multiplicity N is set to be equal to the order of nominal plant n, since $N \geq n$ is the necessary condition for perfect tracking [23]. This assumption can be relaxed to deal with more general systems where $N \neq n$ by using the formulation described in [23].

Consider the continuous-time n^{th} order plant described by

$$\dot{\mathbf{x}}(t) = \mathbf{A}_c\mathbf{x}(t) + \mathbf{b}_cu(t), \quad p(t) = \mathbf{c}_c\mathbf{x}(t). \tag{5.58}$$

The discrete-time state-space equation discretized using the shorter sampling period T_u is given by

$$\mathbf{x}[k+1] = \mathbf{A}_s\mathbf{x}[k] + \mathbf{b}_su[k], \tag{5.59}$$

where $\mathbf{x}[k] = \mathbf{x}(kT_u)$ and

$$\mathbf{A}_s := e^{\mathbf{A}_cT_u}, \quad \mathbf{b}_s := \int_0^{T_u} e^{\mathbf{A}_c\tau}\mathbf{b}_cd\tau. \tag{5.60}$$

By calculating the state transition from $t = iT_y = kT_u$ to $t = (i+1)T_y = (k+n)T_u$, as shown in Figure 5.34, the discrete-time plant $P[z]$ can be represented by

$$\mathbf{x}[i+1] = \mathbf{A}\mathbf{x}[i] + \mathbf{B}\mathbf{u}[i], \quad p[i] = \mathbf{c}\mathbf{x}[i], \tag{5.61}$$

where $\mathbf{x}[i] = \mathbf{x}(iT_y)$, $z := e^{sT_y}$, and multi-rate input vector $\mathbf{u}$ is defined in the lifted form as

$$\begin{aligned}
\mathbf{u}[i] &:= [u_1[i], \cdots, u_n[i]]^T \\
&= [u(kT_u), \cdots, u((k+n-1)T_u)]^T, \tag{5.62}
\end{aligned}$$

where $u_j[i]$ is the inter-sample control input, as shown in Figure 5.34. The state matrices in (5.61) are given by

$$\mathbf{A} = \mathbf{A}_s^n, \mathbf{B} = [\mathbf{A}_s^{n-1}\mathbf{b}_s, \mathbf{A}_s^{n-2}\mathbf{b}_s, \cdots, \mathbf{A}_s\mathbf{b}_s, \mathbf{b}_s], \tag{5.63}$$

$$\mathbf{c} = \mathbf{c}_c. \tag{5.64}$$

5.4.2.2 Design of PTC

The Inter-sample Disturbance Rejection (IDR) was proposed to cancel high order Repeatable Run-Out (RRO) in [20] by modeling the periodic disturbances as Fourier series and the unknown amplitude and phase were estimated by an observer. The IDR method is very effective in cancelling several disturbance modes when the number of selected modes is small. However, the online computational cost for the observer is not negligible when the number of selected modes becomes large. In this section, a novel repetitive control is introduced based on PTC [23] using a Periodic Signal Generator (PSG) [24]. As the PSG can be easily constructed using a series of delay elements z^{-1}, the computational cost is greatly reduced.

First, the PTC is designed using multi-rate feedforward control in the inner loop to obtain the desired command response. The measured output $y[i]$ is assumed to consist of the output disturbance $d[i]$ as

$$y[i] = p[i] - d[i] := \mathbf{c}\,\mathbf{x}[i] - d[i], \tag{5.65}$$

where $p[i]$ is the plant output sampled at a period of T_y. It is worth noting that $p(t)$ is the head position, $d(t)$ is the track runout, and $y(t)$ is the position error in HDD applications. In this section, the disturbance is assumed to be a repetitive signal with a period T_d. From (5.61), the transfer function from the plant state $\mathbf{x}[i+1] \in \mathbf{R}^n$ to the multirate input $\mathbf{u}[i] \in \mathbf{R}^n$ can be derived as

$$\mathbf{u}[i] = \mathbf{B}^{-1}(\mathbf{I} - z^{-1}\mathbf{A})\mathbf{x}[i+1] \tag{5.66}$$

$$= \left[\begin{array}{c|c} \mathbf{O} & -\mathbf{A} \\ \hline \mathbf{B}^{-1} & \mathbf{B}^{-1} \end{array}\right]\mathbf{x}[i+1]. \tag{5.67}$$

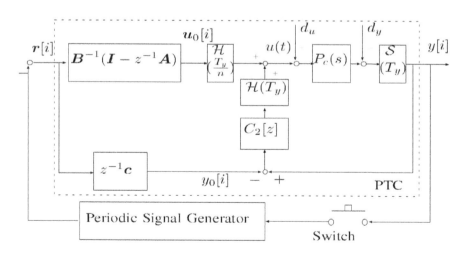

FIGURE 5.35: Block diagram of RPTC.

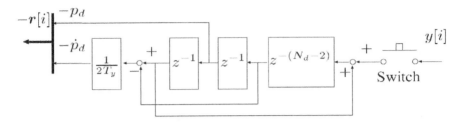

FIGURE 5.36: PSG for a second-order system.

The non-singularity of matrix $\mathbf{B}$ is guaranteed for a controllable plant from the definition in (5.63). From (5.67), all the poles of the transfer function (5.66) are zero and (5.66) is a stable inverse system. If the control input is calculated by (5.68), as shown in Figure 5.35, perfect tracking is guaranteed at the sampling points for the nominal system as

$$\mathbf{u}_0[i] = \mathbf{B}^{-1}(\mathbf{I} - z^{-1}\mathbf{A})\mathbf{r}[i] \tag{5.68}$$

is the *exact* plant inverse [23]. $\mathbf{r}[i](:= \mathbf{x}_d[i + 1])$ is the previewed desired trajectory of the plant state and the nominal output can be calculated as

$$y_0[i] = \mathbf{c}\,\mathbf{x}_d[i] = z^{-1}\mathbf{c}\,\mathbf{r}[i]. \tag{5.69}$$

When the tracking error $y[i] - y_0[i]$ is caused by unmodeled disturbance or modeling error, it can be attenuated by the the robust feedback controller $C_2[z]$, as shown in Figure 5.35.

5.4.3 Design of RPTC

In this section, the PSG is designed to generate the desired trajectory $\mathbf{r}[i]$. As perfect tracking when $\mathbf{x}[i] = \mathbf{x}_d[i]$ or $\mathbf{x}[i] = z^{-1}\mathbf{r}[i]$ is assured, the nominal inner-loop system is expressed as

$$y[i] = z^{-1}r[i] - d_2[i], \ r[i] := \mathbf{c}\,\mathbf{r}[i], \tag{5.70}$$

where $d_2[i] := (1 - P[z]C_2[z])^{-1}d[i]$ and $P[z]$ is the single-rate plant with a sampling period of T_y if the inner-loop feedback controller $C_2[z]$ is of single rate. In RPTC, two schemes can be considered, namely, the *feedback* and *feedforward* approaches. The switch in Figure 5.35 is always on when using the feedback scheme or FeedBack RPTC (FB-RPTC), and the PSG can be designed as the outer-loop controller with

$$r[i] = -\frac{z}{z^{N_d} - 1}y[i], \tag{5.71}$$

where the integer N_d is defined as $\frac{T_d}{T_y}$. From (5.70) and (5.71), the overall closed-loop system is represented by

$$y[i] = -\frac{z^{N_d} - 1}{z^{N_d}}d_2[i]. \tag{5.72}$$

As such, the repetitive disturbance modeled by $d[i] = (z^{N_d} - 1)^{-1}$ is *completely* rejected at every sampling point in steady-state.

In (5.70), redundancy exists when choosing $\mathbf{r}[i] \in \mathbf{R}^n$ from the PSG output $r[i]$ since we have the freedom to select the state variable $\mathbf{x}$. In order to smooth the multi-rate input, the derivative form $\mathbf{x} = [p, \dot{p}, \ddot{p}, \cdots]^T$ should be used. An example of the second order plant with $\mathbf{x} = [p, \dot{p}]^T$ where the velocity command is generated by $\dot{p}_d[i] = \frac{p_d[i+1] - p_d[i-1]}{2T_y}$ is shown in Figure 5.36.

However, the internal model depicted in (5.71) deteriorates the closed-loop characteristics such as robust stability since the gain of the PSG becomes infinite at high order harmonics of periodic disturbances. This motivates the introduction of the FeedForward RPTC (FF-RPTC) using a switching mechanism.

A simple simulation result to explain the FF-PTC is shown in Figure 5.37. The single sinusoidal disturbance signal of 70 Hz with 20 nm in amplitude is injected, and the rest of the simulation conditions are identical as those in the next section.

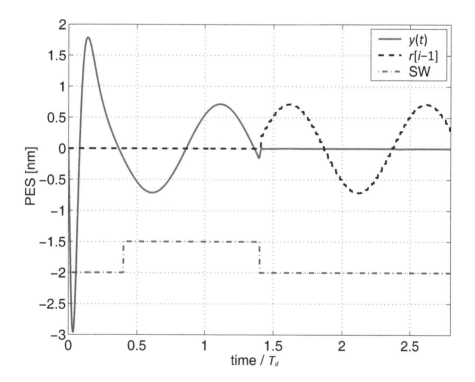

FIGURE 5.37: FF-RPTC algorithm.

The disturbance is injected at $t = 0$ when the switch in Figure 5.35 is in the off state. After the transients in the inner loop with $C_2[z]$ and $P_c(s)$ are settled down, the measured output $y[i]$ is the steady-state response. The switch is then turned to the on state to store the output during one disturbance period T_d, and the switch is turned to and remains in the off state. By using the stored signal, the PSG can reproduce the feedforward signal $\mathbf{r}[i]$ expressed in (5.71) as long as the disturbance is periodic. $y_0[i] (= r[i-1])$ is represented by the dashed line in Figure 5.37. In this case, the PTC generates a control input $u_0[i]$ to cancel the periodic steady-state error. As such, the plant output $p[i]$ perfectly tracks the periodic disturbance $d[i]$ and the tracking error becomes zero at every sampling point $(y[i] = 0)$.

Since the switch turns on for just N_d sampling steps and the PSG has an N_d step delay, the N_d memories work as a complete feedforward compensator. Therefore, the disturbance can be rejected at every sampling point without compromising the feedback characteristics. It should be noted that the signal $y_0[i]$ is generated to prevent any additional transients after the switch is turned off.

5.4.4 Applications to RRO Rejection in HDDs

In this section, the FB-RPTC and FF-RPTC techniques are used to reject RRO in HDDs. The effectiveness of both schemes are compared with simulations and analyses of the obtained results.

Two kinds of disturbances which are injected at the plant output should be considered in the track-following mode of HDDs, namely, the Repeatable Run-Out (RRO) and Non-Repeatable Run-Out (NRRO). While RRO is synchronous to the disk rotation, NRRO is asynchronous and random in phase. Although techniques exist to reject the RRO in the low frequency region, e.g., in [24], RRO is hard to be rejected at higher frequencies using conventional techniques only.

The effects of high-order RRO cannot be neglected since the required servo accuracy is drastically increasing every year. As such, multi-rate repetitive controllers were applied to both the FB- and FF-RPTC [20][21].

For our experiments, the plant is a 2.5" prototype HDD with a Track Pitch (TP) of 450 nm. The sampling period of this drive is $T_y = 210.08$ μs, and the control input is changed $N = 2$ times during this period. The nominal plant $P_n(s)$ is simply modeled as a double integrator and the simulation model $P_a(s)$ including the dead-time and gain variations is given by

$$P_n(s) = \frac{k_{pn}}{ms^2}, \quad P_a(s) = \frac{k_p}{ms^2}e^{-Ls}, \qquad (5.73)$$

where $m = 4.7 \times 10^{-4}$ kg and $k_{pn} = 7.8 \times 10^{-2}$ N/V. The rotational frequency of the spindle motor is 70 Hz and the number of sectors $N_d = 68$. The inner-loop feedback controller $C_2[z]$ is designed by the lead-lag compensator with a crossover frequency of 450 Hz or 18.9% of the Nyquist frequency.

The simulation results are shown in Figures 5.38–5.42. The injected disturbance signal is calculated from the approximate inverse of the sensitivity

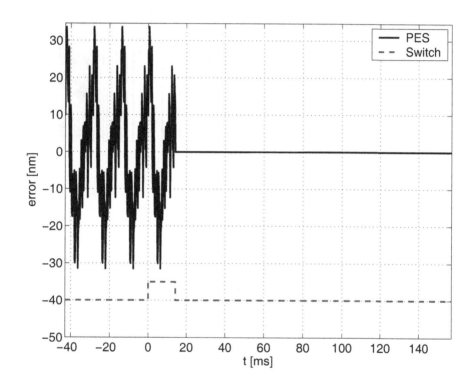

FIGURE 5.38: Simulation of PES using the FF-RPTC and nominal plant with $k_p = 1.0k_{pn}$ and $L = 0$ μs.

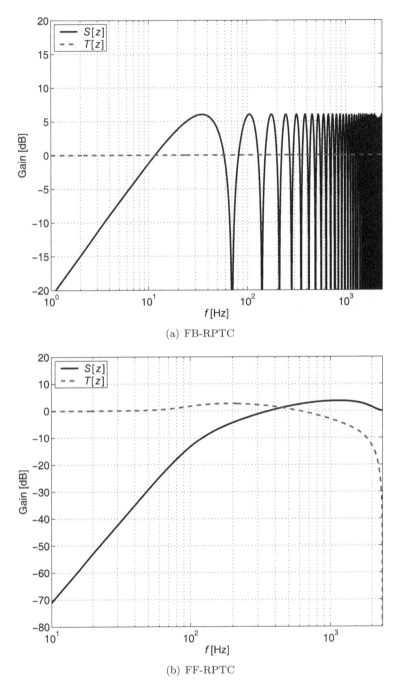

(a) FB-RPTC

(b) FF-RPTC

FIGURE 5.39: Frequency responses of sensitivity transfer functions $S[z]$ and $T[z]$.

transfer function and PES obtained from experiments. As the effects of NRRO are considered in future experiments, only the RRO signal extracted from experimental data by averaging the total PES is considered here. The time response of FF-RPTC for the nominal plant is shown in Figure 5.38. The switch is turned on to start the compensation at the time origin $t = 0$. While the switch is turned on for one disturbance period of $T_d = 14.3$ ms in FF-RPTC only, the switch in the FB–RPTC is on all the time. PTC works from $t = T_d$ to perfectly track the RRO with zero tracking error, as shown in Figure 5.38.

Figures 5.39(a) and 5.39(b) show the sensitivity $S[z]$ and complementary sensitivity $T[z]$ transfer functions of the closed-loop system with the PSG. In FB-RPTC, the sensitivity is zero at the harmonics of 70 Hz since the PSG is the internal model of the periodic disturbance, as can be seen from (5.72). On the other hand, the closed-loop characteristics are determined by the inner-loop with $C_2[z]$ and $P[z]$ since the PSG is a pure feedforward controller in FF-RPTC. As such, a good closed-loop frequency response can be preserved by $C_2[z]$, which can be designed independently.

The case with dead-time $L = 43.26$ μs and a small gain variation of 10% is shown in Figure 5.40. Although a small oscillation is generated, the PES when using the FB-RPTC converges to zero due to the internal model principle. However, the tracking performance is worsened by the plant variation when using the FF-RPTC. To investigate the reason, the FFT analysis of Figure 5.40 is shown in Figure 5.41. Figure 5.41(b) is obtained from the PES after the switch is turned on and Figure 5.41(a) is that without compensation. In the feedforward control, the variation of the command response against plant variations is determined by the sensitivity $(1 - P[z]C_2[z])^{-1}$, which is shown by the solid line in Figure 5.39(a). As such, the performance becomes poor in the high sensitivity band while the RRO is well attenuated in the control bandwidth. In order to overcome this drawback of feedforward control, an adaptive technique was introduced to reduce the modeling error and recover the performance in [25].

The simulation results with a large gain variation of 40% are shown in Figure 5.42. FB-RPTC becomes unstable since the stability robustness is worsened by the PSG, which has an infinite gain at the harmonic frequencies. This can also be understood from the complementary sensitivity transfer function shown in Figure 5.39(a), which shows no roll-off at high frequencies. On the other hand, the FF-RPTC can ensure stability for large plant uncertainties, as shown in Figure 5.42(b).

5.4.5 Experiments on RPTC

In this section, the proposed RPTC methods are verified with experiments. The experiments of FB-RPTC are unstable as they have small stability margins, as stated above. To make it implementable, an LPF is used to eliminate the gain of the peaks of the high order modes to ensure robust stability. This

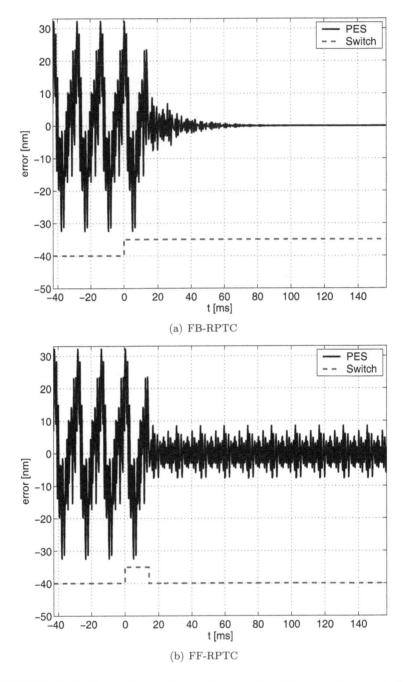

(a) FB-RPTC

(b) FF-RPTC

FIGURE 5.40: Simulation with small variation ($k_p = 1.1k_{pn}$ and $L = 43.26 \ \mu s$).

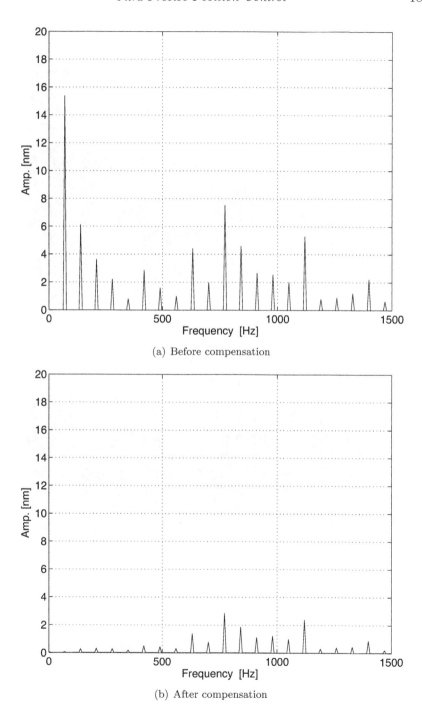

(a) Before compensation

(b) After compensation

FIGURE 5.41: FFT of simulation results.

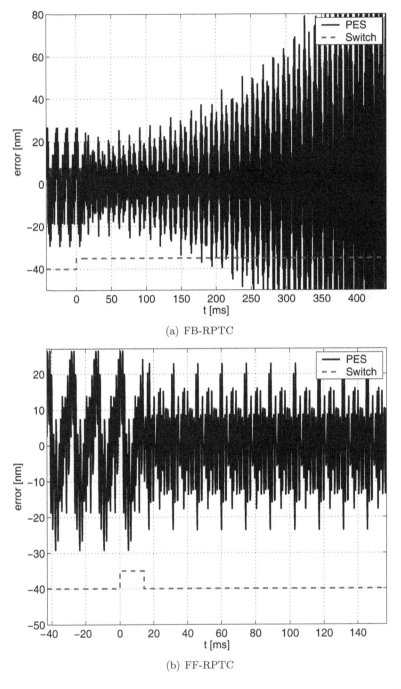

(a) FB-RPTC

(b) FF-RPTC

FIGURE 5.42: Simulation with big variation ($k_p = 1.4k_{pn}$ and $L = 43.26 \ \mu s$).

LPF, called the Q-filter, is implemented as [18]

$$p_f[i] = \frac{z + \gamma + z^{-1}}{\gamma + 2} p_d[i], \tag{5.74}$$

$$\dot{p}_f[i] = \frac{z + \gamma + z^{-1}}{\gamma + 2} \dot{p}_d[i], \tag{5.75}$$

where $\mathbf{r}[i] = [p_f[i], \dot{p}_f[i]]^T$ is utilized instead of $[p_d[i], \dot{p}_d[i]]^T$. Although the Q-filter is improper, it is realizable by making use of the repeatability of $p_d[i+1] = p_d[i+1-N_d]$. A smaller $\gamma(\gamma \geq 2)$ ensures a larger stability margin and roll-off against sensor noise, although the disturbance rejection performance becomes poorer [18]. The frequency response of the Q-filter is shown in Figure 5.43. It is worth noting that the Q-filter is not required in the proposed FF-RPTC due to the switching mechanism.

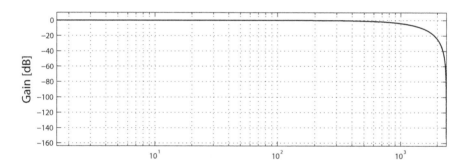

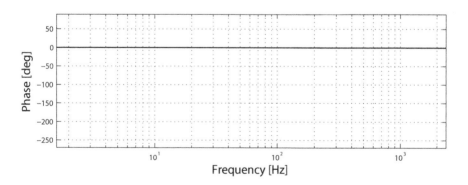

FIGURE 5.43: Frequency response of Q-filter with $\gamma = 2$.

Experimental results of both the FB- and FF-RPTC are shown in Figures 5.44–5.46. In the FF-RPTC, the RRO signals are averaged with respect to the sector number and stored in the PSG. Figure 5.44(a) shows the FFT spectrum of RRO signals with the lead-lag compensator as $C_2[z]$ only. Figures 5.44(b) and 5.44(c) show the FFT spectra of RRO signals with FB- and

FF-RPTCs, respectively. From Figure 5.44(b), it can be observed that the RRO components of the FB-RPTC are almost zero due to the internal model principle. On the other hand, the proposed FF-RPTC has a small PES which is caused by the modeling error of the plant. This is the main disadvantage of the feedforward approach.

NRRO components of FB-RPTC shown in Figure 5.45(b) are greatly amplified as compared to the original $C_2[z]$ in Figure 5.45(a). This is because the internal model worsens the total sensitivity transfer function in NRRO frequencies due to Bode's Integral Theorem, as shown earlier in Figure 5.39(a). Since the FF-RPTC overcomes this problem by the switching mechanism, it can be seen from Figure 5.45(c) that the response is almost identical to that in Figure 5.45(a). The total tracking accuracy is improved by 32.6% when using the proposed switching methods as $3\sigma = 25.8$ nm, which is 5.72% of TP, although the accuracy of the FB-RPTC is worsened by 19.8% as compared to the original $C_2[z]$, as can be seen in Figure 5.46.

In all, two switching-based repetitive controllers named FB-RPTC and FF-RPTC were introduced to reject high order repetitive disturbances along with their advantages and disadvantages. Using the internal model principle, the FB-RPTC approach can guarantee the convergence to zero tracking error against the small plant variations as long as robust stability is ensured. However, a Q-filter is required to ensure closed-loop stability during implementations, which amplifies the NRRO since the total sensitivity transfer function is worsened by the internal model. These disadvantages are present not only in FB-RPTC but also for all the conventional repetitive controllers using an ODOF control structure. On the other hand, the FF-RPTC is a feedforward repetitive compensation controller which ensures robust stability using the switching mechanism. The unavoidable trade-off is that the performance of robust periodic disturbance rejection is worsened in the minor loop. As such, the FB-RPTC without the Q-filter cannot be applied to physical systems while the FF-RPTC possesses good repetitive vibration suppression capabilities.

5.5 Acceleration Feedforward Control (AFC)

In this section, the usage of additional accelerometers for *Acceleration Feedforward Control* (AFC) in HDD head-position control systems is discussed in detail.

5.5.1 Introduction

As mentioned in the previous chapters, HDDs are mass storage devices with a magnetic Read/Write (R/W) head flying above and reading/writing data on disks that are rotating at a high velocity. Due to the structural design

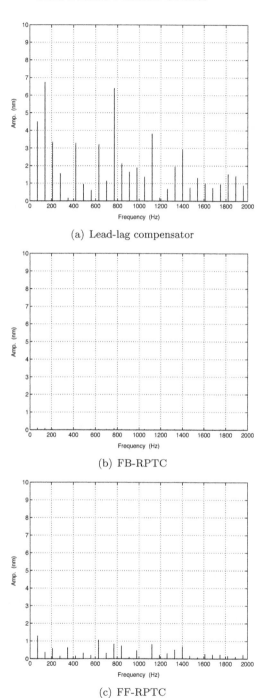

(a) Lead-lag compensator

(b) FB-RPTC

(c) FF-RPTC

FIGURE 5.44: FFT spectra of RRO signals.

194

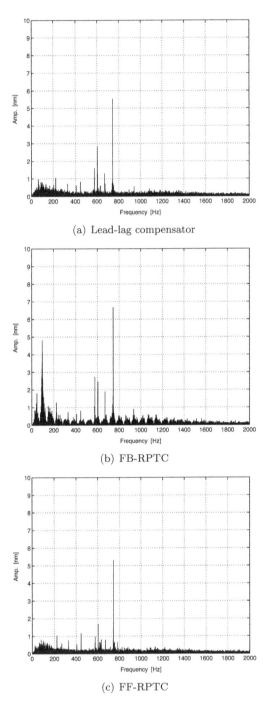

(a) Lead-lag compensator

(b) FB-RPTC

(c) FF-RPTC

FIGURE 5.45: FFT spectra of NRRO signals.

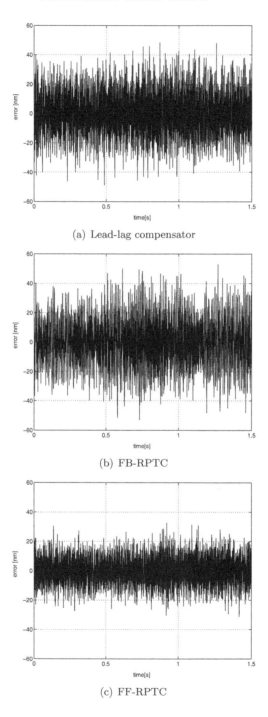

(a) Lead-lag compensator

(b) FB-RPTC

(c) FF-RPTC

FIGURE 5.46: Time responses.

in HDDs, there are situations where the R/W head contacts the disk surface as a result of a large shock or vibration, resulting in data damage or even breakage of the HDD in the worst case scenario. In addition, HDDs are also used in mobile devices such as notebook computers, portable music players, digital video cameras, etc., these days. As such, HDDs are now expected to be used in even harsher environments. Motivated by these problems, current HDDs now have the capability to park the R/W heads away from the disks whenever a large shock or vibration is detected by the acceleration sensor.

In addition, acceleration sensors are also used for improving the performance of the head-positioning control systems in the presence of vibration via Acceleration Feedforward Control (AFC) techniques. This method is made possible with the manufacturing of small, cheap, and high-performance acceleration sensors as a result of advances in semiconductor manufacturing technologies. In HDDs, external disturbances degrade the performance of the head-positioning control systems as disturbances are assumed to be uncontrollable. This is especially crucial for HDDs used in mobile devices. To solve this problem, the AFC is proposed for simplicity and effectiveness as compared to redesigning of the closed-loop system. In this section, the principles and strategies of AFC are discussed.

Currently, acceleration sensors are mass-produced by micro-machining technologies and are capable of measuring acceleration in all three axes. An example is the *piezoresistance*-type acceleration sensor where resistance is generated when its beam is distorted. This variation in resistance can then be used to detect acceleration. In the future, it is expected that other types of acceleration sensors will be applied to several other fields such as mobile devices, entertainment devices, and even security devices, etc.

For HDDs to be adopted for other applications, HDD manufacturing companies are also developing HDDs with embedded three-dimensional acceleration sensors. Also, the harsh operating conditions within an HDD can serve as a test platform for these acceleration sensors, which are small and are mounted like integrated semiconductor chips on circuit boards commercially.

5.5.2 Necessity for AFC

Vibrations from the external environment are transmitted to the R/W head of the HDD via the body of the HDD and the arm on which the head is mounted. The transfer function from external vibration to the PES is given by

$$e = \frac{1}{1+CP}r - \frac{P}{1+CP}d_1 - \frac{P}{1+CP}d_2. \tag{5.76}$$

When there is no external disturbance, the feedback controller is designed such that the transfer function from the reference to the output $\frac{e}{r}$ should be unity in the low frequency range. With the presence of disturbance d_1, the transfer function from the disturbance to the output $\frac{e}{d_1}$ should be zero in the low frequency range. However, if there is an external disturbance d_2 in the high frequency range, the transfer function $\frac{e}{d_2}$ needs to have its gain reduced in the high frequency range, which is identical to the transfer function $\frac{e}{d_1}$. To design a feedback controller to suppress disturbance in the high frequency range, it is necessary to use a smaller sampling period to enhance disturbance suppression capabilities. Unfortunately, this strategy requires additional PES samples from the disk, which decreases the capacity of HDDs and it is also possible that several mechanical resonant modes in the high frequency range are excited. As such, the AFC controller shown in Figure 5.47 is considered to suppress external disturbances on top of that using a feedback controller. The transfer function from disturbances to the PES when AFC is applied is

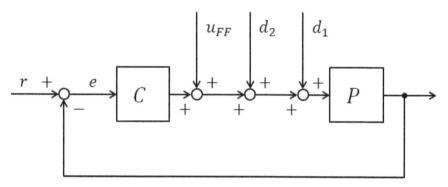

FIGURE 5.47: Block diagram with feedforward input.

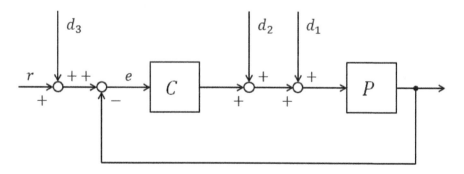

FIGURE 5.48: Block diagram of an HDD subjected to external vibrations.

$$e = \frac{1}{1+CP}r - \frac{P}{1+CP}d_1 - \frac{P}{1+CP}d_2 + \frac{P}{1+CP}u_{FF}. \quad (5.77)$$

If the feedforward input u_{FF} is equal to the external disturbance d_2 ideally, the head-positioning control system will have the same performance as that

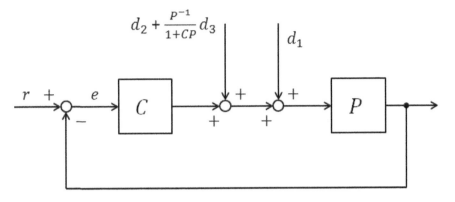

FIGURE 5.49: Block diagram of an HDD considering torque disturbance.

when there are no external disturbances. The low frequency disturbances will be suppressed by the conventional feedback controller while the external disturbances are suppressed by the feedfoward input. Thus, the effects of external disturbances can be eliminated without redesigning the feedback characteristics. The two reasons why AFC is effective in rejecting external disturbances are

- External disturbances are in the high frequency range and cannot be fully suppressed by the conventional feedback controller; and

- External disturbances can be observed and measured by an acceleration sensor.

In this section, the effects of external disturbance are discussed. When an HDD is subjected to external vibrations, the disturbances displace both the head and disk of the HDD and these effects are captured in the block diagram shown in Figure 5.48 (by introducing d_3). It should be noted the the effects of such disturbances differ from that with torque disturbance and should be taken into consideration separately. The block diagram considering torque disturbances is shown in Figure 5.49.

5.5.3 Types of AFC

Recently, many AFC methods have been proposed and the usage of an additional acceleration sensor for vibration suppression control in HDDs had already appeared in 1977 [26]. The four main types of AFC are:

1. Constant-type AFC;

2. Filter-type AFC;

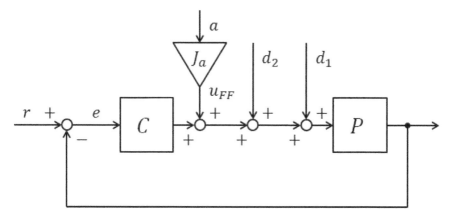

FIGURE 5.50: Block diagram of constant-type AFC.

3. Transfer function-type AFC; and

4. Adaptive identification-type AFC.

5.5.3.1   Constant-Type AFC

First, the effects of external disturbances on the R/W head are considered. In the ideal case where the head is mounted onto the arm, which is assumed to be a rigid body, and the center of rotation is aligned with the center of gravity, external disturbances can be expressed as

$$d_2 = J_a \ddot{\theta}, \tag{5.78}$$

where J_a is the moment of inertia and θ is the rotational angle. The block diagram of the constant-type AFC is shown in Figure 5.50.

If there is a one-axis angular acceleration sensor to measure the angular acceleration of the arm in HDDs, AFC can be readily realized. However, as the center of rotation and the center of gravity are not perfectly aligned in practical situations, the feedforward input in Figure 5.50 is designed as [27]

$$u_{FF} = J_a \ddot{\theta} + a_x \cos(\theta) M \Delta + a_y \sin(\theta) M \Delta, \tag{5.79}$$

where M is the mass of the arm, a_x is the acceleration in the vertical direction, a_y is the acceleration in the horizontal direction, and Δ is the distance between the center of rotation and the center of gravity. With this feedforward input, the external disturbance is eliminated effectively and the torque disturbance arising from the imbalance of the mass of the arm and the center of gravity is also compensated. In this case, it is necessary to use two additional acceleration sensors in addition to the angular acceleration sensor.

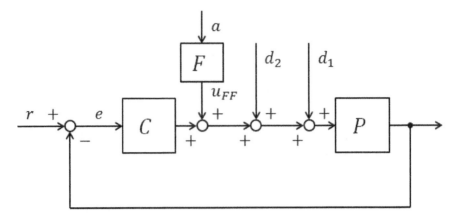

FIGURE 5.51: Block diagram of filter-type AFC with PCF.

5.5.3.2 Filter-Type AFC

It is theoretically possible to suppress all external disturbances using the constant-type AFC. However, the acceleration signal is affected by factors such as computational delay, noise, etc. To overcome this problem, several methods have been proposed to match the feedforward input to the external disturbances through the use of add-on filters [28][29][30].

In [30], the Phase Compensating Filter (PCF) is proposed to match the gain and phase of the feedforward input to the external disturbance in the frequency range from 200 Hz to 500 Hz, as shown in Figure 5.51. This technique can be readily adapted to AFC effectively.

5.5.3.3 Transfer Function-Type AFC

Recently, the transfer functions from the acceleration sensor to the R/W head were used to identify the external disturbances exactly in [31][32].

In [31], a feedforward controller is designed by calculating the transfer function from the output of the acceleration sensor to the disturbance d in order to calculate the AFC from the measurable transfer functions. The transfer function G_{w0} from the output of the acceleration sensor to the PES is measured using

$$
\begin{aligned}
\frac{e}{\ddot{\theta}} &= G_{w0} \\
&= G\frac{-P}{1+PC},
\end{aligned}
$$
(5.80)

and is shown in Figure 5.52.

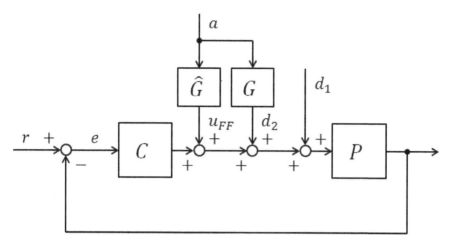

FIGURE 5.52: Block diagram of transfer function-type AFC for transfer function calculation.

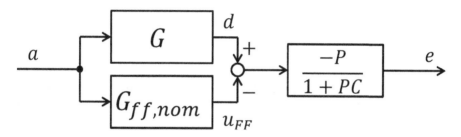

FIGURE 5.53: Block diagram for transfer function calculation using transfer function-type AFC.

Further, the transfer function G_w

$$\frac{e}{\bar{\theta}} = G_w$$

$$= (G - G_{ff,nom})\frac{-P}{1 + PC} \tag{5.81}$$

is measured using the temporary feedforward controller $G_{ff,nom}$ as shown in Figure 5.53.

With the transfer functions G_{w0} and G_w and feedforward controller $G_{ff,nom}$, the transfer functions of the closed-loop transfer function can be calculated from

$$\frac{-P}{1 + PC} = \frac{G - G_{ff,nom}}{G_{ff,nom}}. \tag{5.82}$$

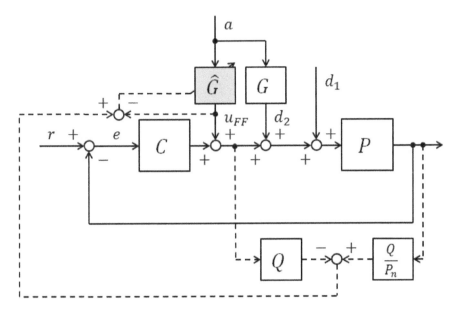

FIGURE 5.54: Block diagram of adaptive identification-type AFC.

Using this transfer function, the transfer function from the acceleration sensor to the disturbance d can now be calculated by dividing by G_{w0}. This transfer function can be used to eliminate the external disturbance effectively.

5.5.3.4 Adaptive Identification-Type AFC

It has also been proposed that an accurate AFC can be calculated using adaptive identification. In [33], the transfer function from the output of the acceleration sensor to the disturbance is identified directly. Using this method, the feedforward controller can track the frequency and gain variations of the external disturbances using adaptive schemes. Also, there is no need to consider external factors such as delay or noise, as this method identifies the transfer functions directly. The feedfoward controller, which includes computing delay or noise elimination considered in the filter-type feedforward controller, is also auto-tuned since this method identifies the transfer function directly. In this section, the adaptive identification-type AFC using Recursive Least Squares (RLS) and gradient methods is introduced for real time identification [34] and their block diagrams are shown in Figures 5.54 and 5.55, respectively.

(a) RLS

In this method, an Infinite Impulse Response (IIR) is used as the feedforward controller and its parameters are identified. The feedforward input is expressed

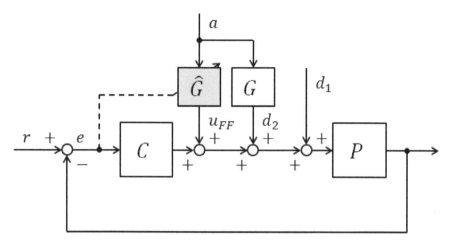

FIGURE 5.55: Block diagram of adaptive identification-type AFC using RLS and gradient method.

as

$$d = \frac{B[z^{-1}]}{A[z^{-1}]}a, \tag{5.83}$$

$$A[z^{-1}] = 1 + a_1 z^{-1} + a_2 z^{-2} + \cdots + a_{N_a} z^{-N_a}, \tag{5.84}$$

$$B[z^{-1}] = 1 + b_1 z^{-1} + b_2 z^{-2} + \cdots + b_{N_b} z^{-N_b}, \tag{5.85}$$

where N_a and N_b are orders of denominator and numerator of the identified transfer function, respectively. As the external disturbances cannot be observed directly, an estimate of the disturbance using the disturbance observer is given by

$$\hat{d} = (-\frac{CP}{1+CP}Q + \frac{Q}{P_n}\frac{P}{1+CP})d - \frac{Q}{P_n}\frac{1}{1+CP}\xi$$

$$= \frac{P/P_n}{1+CP}Q \cdot d + \xi' \cong Q \cdot d + \xi'. \tag{5.86}$$

The parameters of the feedforward controller to be identified are given by

$$\vartheta = [a_1, a_2, \cdots, a_{N_a}, b_1, b_2, \cdots, b_{N_b}], \tag{5.87}$$

and these parameters can be identified with past input and output signals using

$$\varphi = [-\hat{d}[k-1], -\hat{d}[k-2], \cdots, -\hat{d}[k-N_a],$$

$$Q \cdot a[k-1], Q \cdot a[k-2], \cdots, Q \cdot a[k-N_b]]. \tag{5.88}$$

It is worth nothing that an LPF Q is added to the input signals considering the effects of an LPF on the disturbance observer.

The update equations for the RLS algorithm are

$$\vartheta[k] = \vartheta[k-1] + \frac{\Gamma[k-1]\varphi[k]}{1 + \varphi^T[k]\Gamma[k-1]\varphi[k]}\varepsilon[k], \tag{5.89}$$

$$\varepsilon[k] = \hat{d}[k-1] - \varphi^T[k]\vartheta[k-1], \tag{5.90}$$

$$\Gamma[k] = \Gamma[k-1] - \frac{\Gamma[k-1]\varphi[k]\varphi^T[k]\Gamma[k-1]}{1 + \varphi^T[k]\Gamma[k-1]\varphi[k]}, \tag{5.91}$$

where Γ is a covariance matrix and the parameters of the feedforward controller are identified such that the error in (5.90) is minimized. From this process, the parameters are determined and the feedforward input is calculated by

$$u_{FF} = \frac{B[z^{-1}]}{A[z^{-1}]}a. \tag{5.92}$$

(b) Gradient method

The adaptive identification using the RLS method mentioned previously is a general method for online identification in real time. However, divisions are required in the update equations which make the RLS adaptive identification method computationally intensive. In this section, a gradient method is proposed which can be readily applied to HDDs. In this method, a feedforward Finite Impulse Response (FIR) is expressed as

$$d = B[z^{-1}]a, \tag{5.93}$$

$$B[z^{-1}] = 1 + b_1 z^{-1} + b_2 z^{-2} + \cdots + b_{N_b} z^{-N_b}, \tag{5.94}$$

$$\vartheta = [b_1, b_2, \cdots, b_{N_b}], \tag{5.95}$$

where N_b is the order of the FIR and φ is a vector of parameters to be identified. Here, PES is used in the error function for the update equations and the parameters of the FIR filter are calculated to minimize the error function given by

$$e = \frac{1}{1 + PC}d + \frac{P}{1 + PC}\sum_{k=0}^{n} b_k z^{-k} u. \tag{5.96}$$

The update equation is given by

$$\theta_i[k+1] = \theta_i[k] - \mu\Delta e[k]^2/\Delta\theta_i. \tag{5.97}$$

Using (5.96), the update equation can be rewritten as

$$\theta_i[k] = \theta_i[k] - \mu 2e[k]\Delta e[k]/\Delta\theta_i \tag{5.98}$$

$$= \theta_i[k] - \mu 2e[k]\frac{P}{1 + PC}u[k-i]. \tag{5.99}$$

From the equations above, it is obvious that the adaptive identification with gradient method is less computationally intensive as compared to the RLS method.

5.5.4 Performance Evaluation for AFC

Many AFC methods have been discussed in the previous sections. In this section, rejection of external disturbances using the method proposed in [34] is tested on an HDD. In the experiment, the HDD is placed on an exciter which subjects the HDD to a 100 Hz vibration. The experimental setup is shown in Figure 5.56.

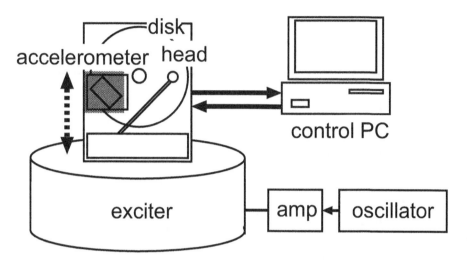

FIGURE 5.56: Experimental setup with shaker for verification of AFC.

The feedforward input is calculated based on the measured acceleration and PES. The experimental results in the time domain are shown in Figure 5.57.

In the time interval from 0 to 500 ms, the AFC input was not applied. As such, the PES has large oscillations. When the AFC input was applied from 500 to 2000 ms, the amplitude of oscillation of PES is reduced drastically. The frequency spectra of the PES with and without AFC are shown in Figures 5.58 and 5.59. With the application of AFC, the 100 Hz vibrational disturbance can be eliminated without any negative impacts on the other frequencies.

5.5.5 Applications of AFC

AFC has already been applied to HDDs. However, this control technique can also be readily applied to other fields which require servo control in the presence of external vibrations. In this section, many types of AFC have been proposed, and the control systems designer can easily select a method which is suitable for his field of application considering costs, hardware specifications, performance, etc.

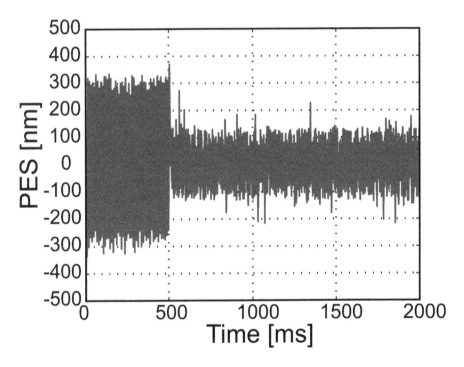

FIGURE 5.57: Time series of PES with and without proposed AFC.

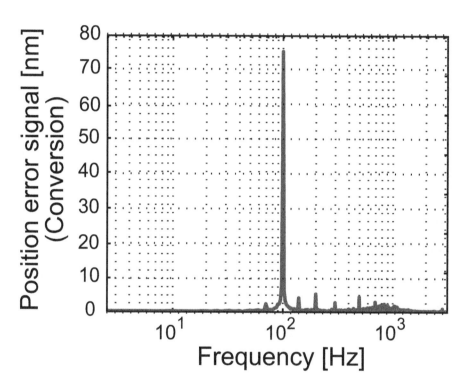

FIGURE 5.58: Frequency spectrum of PES without AFC.

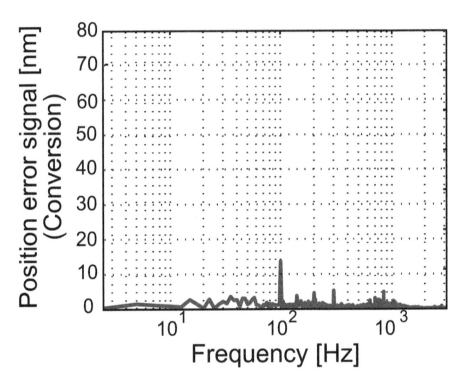

FIGURE 5.59: Frequency spectrum of PES with AFC.

5.5.5.1 Application to Vehicles

Acceleration sensors are currently mounted in many vehicles. An example is the use of acceleration sensors in the air bag system. In addition, acceleration sensors can also be used to improve other aspects of the vehicles, e.g., including AFC in the vehicular control system to improve the ride comfort and to reduce the chassis vibrations, etc.

5.5.5.2 Application to Industrial Robots

Currently, industrial robots are mainly controlled using high gain feedback along each DOF. However, the demand for precision has become one of the most important aspects in engineering recently. In this area, vibration suppression is an important problem, and this problem can be easily solved using additional sensors and the proposed AFC.

Bibliography

[1] T. Atsumi, T. Arisaka, T. Shimizu, and T. Yamaguchi, "Vibration Servo Control Design for Mechanical Resonant Modes of a Hard-Disk-Drive Actuator," *JSME International Journal Series C*, Vol. 46, No. 3, pp. 819–827, 2003.

[2] G. F. Franklin, J. D. Powell, and M. Workman, *Digital Control of Dynamic Systems*, Addison Wesley Longman, 1998.

[3] K. Zhou, J. C. Doyle, K. Glover, *Robust Optimal Control*, Prentice-Hall, 1996.

[4] J. C. Doyle, B. A. Francis, and A. R. Tannenbaum, *Feedback Control Theory*, Macmillan Publishing Company, 1992.

[5] B. A. Francis, *A Course in H_∞ Control Theory*, Springer Verlag, New York, 1987.

[6] T. Mita, H_∞ *Control*, Shoukoudou, 1994 (in Japanese).

[7] K. Nonami, H. Nishimura, and M. Hirata, *Control System Design Using MATLAB*, TDU Press, 1998 (in Japanese).

[8] K. Glover and J. C. Doyle, "State-Space Formulae for All Stabilizing Controllers that Satisfy and H_∞-Norm Bound and Relations to Risk Sensitivity," *Systems & Control Letters*, Vol. 11, pp. 167–172, 1988.

[9] T. Iwasaki, *LMI and Control*, Shoukoudou, 1997 (in Japanese).

[10] M. Hirata, K-. Z. Liu, and T. Mita, "Head Positioning Control of a Hard Disk Drive Using H_∞ Theory," in *Proceedings of the 31th IEEE Conference on Decision and Control*, Vol. 2, pp. 2460–2461, 1992.

[11] M. Hirata, M. Takiguchi, and K. Nonami, "Track-Following Control of Hard Disk Drives Using Multi-Rate Sampled-Data H_∞ Control," in *Proceedings of the 42nd IEEE Conference on Decision and Control*, pp. 3414–3419, 2003.

[12] T. Chen and B. Francis, *Optimal Sampled-Data Control Systems*, Springer Verlag, 1995.

[13] S. Hara, Y. Yamamoto, and H. Fujioka, *Sampled-Data Control Toolbox Manual*, Cybernet Systems, 2005.

[14] T. Chen and L. Qiu, "H_∞ Design of General Multi-rate Sampled-Data Control Systems," *Automatica*, Vol. 30, No. 7, pp. 1139–1152, 1994.

[15] T. Hara and M. Tomizuka, "Multi-Rate Controller for Hard Disk Drives with Redesign of State Estimator," in *Proceedings of the American Control Conference*, Vol. 5, pp. 3033–3037, 1998.

[16] J. Ishikawa, "A Study on Multirate Sampled-Data Control for Hard Disk Drives," in *The Papers of Technical Meeting on Industrial Instrumentation and Control, IEEJ Japan*, IIC-00-55, pp. 31–38, 2000 (in Japanese).

[17] T. Semba, "An H_∞ Design Method for a Multi-Rate Servo Controller and Applications to a High Density Hard Disk Drive," in *Proceedings of the 40th IEEE Conference on Decision and Control*, pp. 4693–4698, 2001.

[18] K. K. Chew and M. Tomizuka, "Digital Control of Repetitive Errors in Disk Drive Systems," *IEEE Control Systems Magazine*, Vol. 10, No. 1, pp. 16–20, 1990.

[19] S. Hara, Y. Yamamoto, T. Omata, and M. Nakano, "Repetitive Control System—A New-Type Servo System," *IEEE Transactions on Automatic Control*, Vol. 33, pp. 659–668, 1988.

[20] H. Fujimoto, F. Kawakami, and S. Kondo, "Multirate Repetitive Control and Applications—Verification of Switching Scheme by HDD and Visual Servoing," in *Proceedings of the American Control Conference*, pp. 2875–2880, 2003.

[21] H. Fujimoto, "RRO Compensation of Hard Disk Drives With Multirate Repetitive Perfect Tracking Control," *IEEE Transactions on Industrial Electronics*, Vol. 56, No. 10, pp. 3825–3831, 2009.

[22] M. Tomizuka, "Zero Phase Error Tracking Algorithm For Digital Control," *Transactions of the ASME, Journal of Dynamic Systems, Measurement, and Control*, Vol. 109, pp. 65–68, March 1987.

[23] H. Fujimoto, Y. Hori, and A. Kawamura, "Perfect Tracking Control Based on Multirate Feedforward Control with Generalized Sampling Periods," *IEEE Transactions on Industrial Electronics*, Vol. 48, No. 3, pp. 636–644, 2001.

[24] C. Kempf, W. Messner, M. Tomizuka, and R. Horowitz, "Comparison of Four Discrete-Time Repetitive Algorithms," *IEEE Control Systems Magazine*, Vol. 13, No. 5, pp. 48–54, 1993.

[25] H. Fujimoto, F. Kawakami, and S. Kondo, "Repetitive Control of Hard Disk Drive Based on Switching Scheme and Gain Adaptation," in *Proceedings of the IEE of Japan Technical Meeting*, No. IIC-03-92, pp. 1–6, 2003 (in Japanese).

[26] R. A. White, *Shock Force Compensating System*, US Patent 4040103, 1977.

[27] R. Oboe, "Use of MEMS Based Accelerometers in Hard Disk Drives," in *Proceedings of the 2001 IEEE/ASME International Conference on Advanced Intelligent Mechatronics*, pp. 1142–1147, 2001.

[28] D. Y. Abramovitch, "Rejecting Rotational Disturbance on Small Disk Drives Using Rotational Accelerometers," *Control Eng. Practice*, Vol. 5, No. 11, 1517–1524, 1997.

[29] S-. E. Baek, S-. H. Lee, "Vibration Rejection Control for Disk Drives by Acceleration Feedforward Control," in *Proceedings of the 38th Conference on Decision & Control*, pp. 5259–5262, Phoenix, AZ , USA, December 7–10, 1999.

[30] A. Jinzenji, T. Sasamoto, K. Aikawa, S. Yoshida, and K. Aruga, "Acceleration Feedfoward Control Against Rotational Disturbance in Hard Disk Drives," *IEEE Transaction on Magnetics*, Vol. 37, No. 2, pp. 888–893, 2001.

[31] M. T. White and M. Tomizuka, "Increased Disturbance Rejection in Magnetic Disk Drives by Acceleration Feedfoward Control," in *Proceedings of the 1996 IFAC*, pp. 489–494, 1996.

[32] S. Pannu and R. Horowitz, "Accelerometer Feedforward Servo for Disk Drives," in *Proceedings of the IEEE International Conference on Advanced Intelligent Mechatronics*, 1997.

[33] K. Usui, M. Kisaka, A. Okuyama, and M. Nagashima, "Reduction of External Vibration in Hard Disk Drives Using Adaptive Feedforward Control with a Single Shock Sensor," in *Proceedings of the 9th International Workshop on Advanced Motion Control*, 2006.

[34] N. Bando, S. Oh, and Y. Hori, "Disturbance Rejection Control on Adaptive Identification of Transfer Characteristics from Acceleration Sensor for Hard Disk Drives System," *IEEJ Transactions on Industry Applications*, Vol. 123, No. 12, pp. 1461–1466, 2003 (in Japanese).

Chapter 6

Control Design for Consumer Electronics

Mitsuo Hirata

Utsunomiya University

Shinji Takakura

Toshiba Corp.

Atsushi Okuyama

Tokai University

6.1   Control System Design for Energy Efficiency

Apart from optical disks such as Compact Discs (CDs) or Mini Discs (MDs), Hard Disk Drives (HDDs) have been used as recording devices for portable music players in recent years. For such devices, it is better to reduce the power consumption as much as possible in order to achieve a long battery life or a small battery size. Thus, various attempts have been made for energy saving. For example, the power of the spindle motor is cut off when there is no data access, or the order of recording data is sorted so that unnecessary track-seeking responses do not occur. In this section, recent work to reduce the power consumption from the control theory view point is introduced.

The processors (in which control algorithms are implemented) generally have an energy saving mode, and the power consumption of the processor is reduced when there are no computation activities. As such, power consumption may be reduced indirectly if the amount of calculations can be reduced. When only low-end processors are available due to cost or size restriction, the amount of the controller calculations is also constrained.

The easiest way to reduce the amount of computation load is through controller order reduction. Obtaining a minimum realization where uncontrollable or unobservable modes are removed, performing a balanced-order truncation, or using the Hankel norm approximation are well-known approaches to reduce the order of the controller. However, these methods do not always work well, and it generally depends on how much the order of the controller can be reduced. In this section, other approaches to reduce the amount of controller calculations are introduced.

6.1.1 Interlacing Controller

In general, a digital controller consists of both slow and fast modes. For example, the integrator and differentiator respond to slow modes and fast modes in a Proportional-Integral-Derivative (PID) controller, respectively. Since the slow mode responds to slow-changing signals, it can be updated less frequently in order to reduce the computational load. Hence, the resulting multi-rate digital controller consists of a fast controller at a high sampling rate and a slow controller at a low sampling rate. To ensure that the amount of calculations can be uniformly reduced at each sampling period, an interlacing technique is proposed. In this section, the details of the interlacing controller are described.

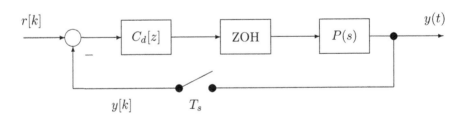

FIGURE 6.1: Track-following control system.

The nominal plant model $P(s)$ of the track-following control system shown in Figure 6.1 is assumed to be a double integrator plant

$$P(s) = \frac{1}{mT_p} \frac{1}{s^2},$$

(6.1)

where $m = 0.001$ kg and $T_p = 2.54 \times 10^{-7}$ m/track. For this system, a PID controller was designed as

$$C(s) = 350.7 \frac{(s + 1143)(s + 628)}{s(s + 4.3 \times 10^4)(s + 8631)}.$$

(6.2)

In order to implement this as a digital controller, $C(s)$ was discretized using Tustin transformation at a sampling period of $T_s = 37.9$ μs and the corresponding digital controller $C_d[z]$ is given by

$$C_d[z] = 3.25 \times 10^{-3} \frac{(z+1)(z-0.9765)(z-0.9576)}{(z-1)(z-0.719)(z-0.1)}. \tag{6.3}$$

$C_d[z]$ contains two slow modes at 1 and 0.719, and one fast mode at 0.1. Applying partial fraction expansion to (6.3), $C_d[z]$ can be rewritten as

$$C_d[z] = \underbrace{\frac{2.56 \times 10^{-5}}{z-1}}_{C_{ds1}[z]} + \underbrace{\frac{-1.97 \times 10^{-3}}{z-0.719}}_{C_{ds2}[z]} + \underbrace{\frac{4.82 \times 10^{-3}}{z-0.1} + 3.25 \times 10^{-3}}_{C_{df}[z]}. \tag{6.4}$$

The control system using (6.4) is shown in Figure 6.2.

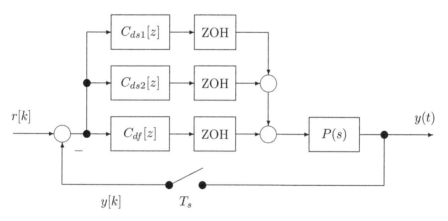

FIGURE 6.2: Parallel representation of controller $C_d[z]$.

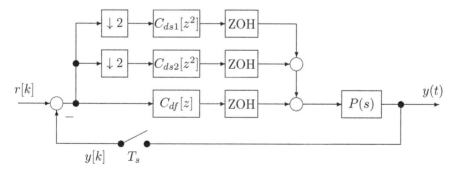

FIGURE 6.3: Parallel representation of controller with down-samplers.

Since $C_{ds1}[z]$ and $C_{ds2}[z]$ contain slow modes, their sampling periods can be chosen to be twice as long. As such, the block diagram in Figure 6.3 can be obtained, where $\downarrow 2$ denotes a down-sampling by a factor of two. We present the following lemma to calculate $C_{ds1}[z^2]$ and $C_{ds2}[z^2]$.

Lemma 1 *A first-order discrete time block is given by $\frac{1}{z-a}$. If its input is held constant for N time steps, the transfer function at a slower sampling rate by a factor of N is*

$$\frac{a^{N-1} + a^{N-2} + \ldots + a + 1}{z^N - a^N}. \tag{6.5}$$

▪

From this lemma, we have

$$C_{ds1}[z^2] = \frac{5.12 \times 10^{-5}}{z^2 - 1}, \quad C_{ds2}[z^2] = \frac{-3.39 \times 10^{-3}}{z^2 - 5.17 \times 10^{-1}}. \tag{6.6}$$

Using the down-sampler, the total amount of computation can be reduced since $C_{ds1}[z^2]$ and $C_{ds2}[z^2]$ are updated only when $k = 2n$, where $n = 0, 1, 2, \ldots$ However, the amount of computation is essentially the same as the original single-rate digital controller at the sampling instants when $k = 2n$. The non-uniform nature of the amount of computations is not attractive from a practical point of view. As such, an alternative method to update $C_{ds1}[z^2]$ and $C_{ds2}[z^2]$ is proposed in [1]. As shown in Figure 6.4, $C_{ds1}[z^2]$ and $C_{ds2}[z^2]$ are updated when $k = 2n$ and $k = 2n+1$, respectively. Therefore, the amount of computations can be reduced not only when $k = 2n + 1$, but also when $k = 2n$. This controller is referred to as the *multi-rate interlacing controller.*

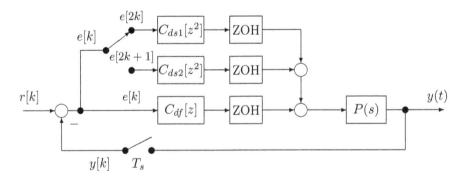

FIGURE 6.4: Multi-rate interlacing controller.

To evaluate the control performance of the multi-rate interlacing controller, white noise is injected at the plant input. The time response of output y is shown in Figure 6.5. The upper, middle, and lower figures show the outputs y using the controller in (6.3), the multi-rate interlacing controller, and the

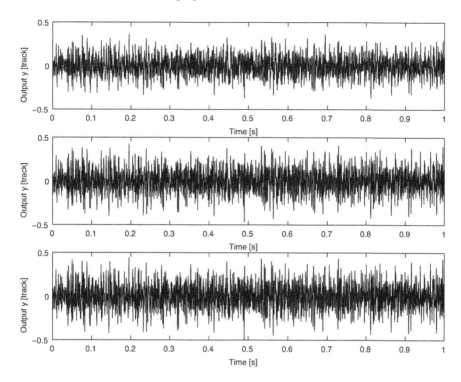

FIGURE 6.5: Comparison of track-following performance.

single-rate controller discretized by Tustin transformation at a sampling period of $2T_s$, respectively. The 3σ values of these outputs are 0.32, 0.39, and 0.41, respectively. From Figure 6.5, it can be seen that multi-rate interlacing controller outperforms the single-rate controller with a sampling period of $2T_s$. Although the performance improvement is less than exemplary in this simulation, it is possible that the multi-rate interlacing controller achieves almost the same performance as the single-rate controller at a faster sampling rate [1].

6.1.2 Short-Track Seeking Using TDOF Control with IVC

During short track-seeking control in HDDs, it is effective to use unified control methods which work for both track-seeking and track-following. Without mode switching, fast and accurate seeking can be achieved because unexpected transient responses during mode switching will not occur. This can be achieved based on the Two-Degrees-of-Freedom (TDOF) control scheme.

In the TDOF control scheme, feedforward controllers are introduced in addition to the feedback controller. As such, the amount of computation will be increased. For small HDDs that are used for notebook computers or portable

218

music players, high-performance microprocessors are not used due to cost
or power consumption limitations. In this section, a method to improve the
short track-seeking response without increasing the amount of computation is
presented [2].

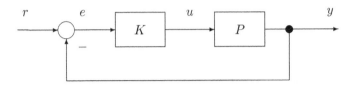

FIGURE 6.6: ODOF control system.

The One-Degree-of-Freedom (ODOF) control system shown in Figure 6.6
cannot improve the reference response from r to y without affecting other
feedback performances such as disturbance rejection and robust stability to
plant uncertainties. However, if the states of the feedback controller can be
set to an arbitrary value at the moment when the step input is injected to
the reference r, the transient response of y will be changed. Therefore, it is
possible to improve the step response by optimizing the initial value of the
feedback controller.

The state-space equations of a discrete-time plant model $P[z]$ are given by

$$x_p[k+1] \quad = \quad A_p x_p[k] + B_p u[k], \tag{6.7}$$
$$y[k] \quad = \quad C_p x_p[k], \tag{6.8}$$

where $P[z]$ is a Single-Input-Single-Output (SISO) system of order n_p. The
state-space equations of a feedback controller $K[z]$ are

$$x_k[k+1] \quad = \quad A_k x_k[k] + B_k(r[k] - y[k]), \tag{6.9}$$
$$u[k] \quad = \quad C_k x_k[k] + D_k(r[k] - y[k]), \tag{6.10}$$

where the order is n_k. From (6.7)–(6.10), the state-space representation of the
closed-loop system in Figure 6.6 is given as

$$x[k+1] \quad = \quad Ax[k] + Br[k], \tag{6.11}$$
$$y[k] \quad = \quad Cx[k], \tag{6.12}$$

where

$$x[k] := \left[\begin{array}{c} x_p[k] \\ x_k[k] \end{array} \right], \quad A := \left[\begin{array}{cc} A_p - B_p D_k C_p & B_p C_k \\ -B_k C_p & A_k \end{array} \right], \tag{6.13}$$

$$B := \left[\begin{array}{c} B_p D_k \\ B_k \end{array} \right], \quad C := \left[\begin{array}{cc} C_p & 0 \end{array} \right]. \tag{6.14}$$

We introduce the following assumption.

Assumption 2

(A1) *The feedback controller K has been designed to meet the desired performance specifications. As such, the closed-loop system shown in Figure 6.6 is stable, i.e., all the eigenvalues of A are located inside of the unit disk.*

(A2) *Reference input r is a step signal.*

(A3) *Initial value of the plant is zero, i.e., $x_p[0] = 0$.*

From **(A1)** and **(A2)**, $x[k]$ converges to a certain value $x[\infty]$ for the step input r as $k \to \infty$. We can define an error-state vector as

$$\begin{bmatrix} e_p[k] \\ e_k[k] \end{bmatrix} := \begin{bmatrix} x_p[k] - x_p[\infty] \\ x_k[k] - x_k[\infty] \end{bmatrix}, \quad e[k] := \begin{bmatrix} e_p[k] \\ e_k[k] \end{bmatrix}. \tag{6.15}$$

Since **(A1)** guarantees the existence of $(I - A)^{-1}$, we have

$$x[\infty] = (I - A)^{-1} Br. \tag{6.16}$$

Therefore, the closed-loop system depicted in (6.11)–(6.12) can be described by the error equation as

$$e[k + 1] = Ae[k]. \tag{6.17}$$

Now, we consider obtaining the initial value of the feedback controller $x_k[0]$ which minimizes the performance index

$$J = \sum_{k=0}^{\infty} e^T[k] Q e[k], \tag{6.18}$$

where $(Q^{\frac{1}{2}}, A)$ is assumed to be observable.

The *discrete-time Lyapunov equation* is given by

$$Q = P - A^T P A, \tag{6.19}$$

and has a positive definite solution $P > 0$. (6.18) can then be rewritten as

$$\begin{aligned} J &= e^T[0] P e[0] \\ &= e_p^T[0] P_{11} e_p[0] + 2 e_p^T[0] P_{12} e_k[0] + e_k^T[0] P_{22} e_k[0], \end{aligned} \tag{6.20}$$

where P is partitioned as

$$P =: \begin{bmatrix} P_{11} & P_{12} \\ P_{12}^T & P_{22} \end{bmatrix} \tag{6.21}$$

to meet the dimensions of $e_p[0]$ and $e_k[0]$. Since P is positive definite, J will be minimum and

$$\frac{\partial J}{\partial e_k[0]} = 2P_{12}^T e_p[0] + 2P_{22} e_k[0] = 0 \tag{6.22}$$

is satisfied. From $P_{22} > 0$, we have

$$e_k[0] = -P_{22}^{-1}P_{12}e_p[0]. \tag{6.23}$$

The initial value of the feedback controller that minimizes J from (A3) and (6.16) can be further expressed as

$$x_k[0] = \begin{bmatrix} P_{22}^{-1}P_{12}^T, & I \end{bmatrix}(I - A)^{-1}Br. \tag{6.24}$$

It is worth noting that the control input is not evaluated in (6.18). Nevertheless, it is possible to evaluate the control input as

$$J = \sum_{k=0}^{\infty} \left\{ e[k]^T Q_e e[k] + q_u (u[k] - u[\infty])^2 \right\}. \tag{6.25}$$

In this case, $u[k]$ can be described as

$$\begin{aligned}
u[k] &= C_k x_k[k] + D_k(r - y[k]) \\
&= -D_k C_p x_p[k] + C_k x_k[k] + D_k r \\
&= \underbrace{[-D_k C_p, \ C_k]}_{C_u} x[k] + D_k r. \tag{6.26}
\end{aligned}$$

Hence, we have

$$u[k] - u[\infty] = C_u e[k], \tag{6.27}$$

and (6.25) can be represented as

$$J = \sum_{k=0}^{\infty} e^T[k]\underbrace{(Q_e + q_u C_u^T C_u)}_{Q} e[k], \tag{6.28}$$

which has the same form as (6.18). Note that when we use the performance index depicted in (6.25), it is easy to show that $(Q^{\frac{1}{2}}, A)$ is observable if both $(Q_e^{\frac{1}{2}}, A)$ and (C_k, A_k) are observable and $Q_e \geq 0$ and $q_u > 0$ are satisfied.

The time responses of the output y and corresponding control input u of the control system with an ODOF controller are shown in Figure 6.7 and are denoted by dashed lines. A step input of $r = 1$ track and unit impulse disturbance are injected at $t = 0$ ms and $t = 5$ ms, respectively. From this figure, it can be observed that the obtained response by the ODOF controller is not satisfactory due to a large overshoot.

To improve the step response, the proposed method is applied. The state-space equation of the double-integrator plant depicted in (6.1) is given by

$$\begin{bmatrix} \dot{x}_1 \\ \dot{x}_2 \end{bmatrix} = \begin{bmatrix} 1 & 1 \\ 0 & 1 \end{bmatrix}\begin{bmatrix} x_1 \\ x_2 \end{bmatrix} + \begin{bmatrix} 0 \\ \frac{1}{m} \end{bmatrix}u, \tag{6.29}$$

$$y = \begin{bmatrix} \frac{1}{T_p} & 0 \end{bmatrix}\begin{bmatrix} x_1 \\ x_2 \end{bmatrix}, \tag{6.30}$$

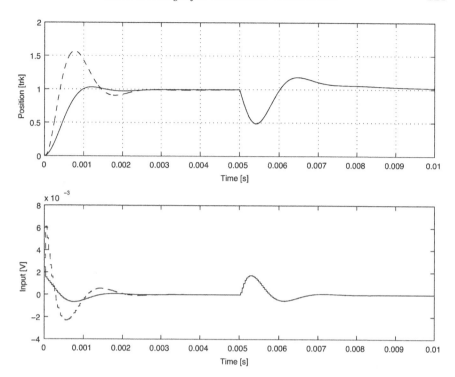

FIGURE 6.7: Improvement of step response by initial value compensation.

where x_1, x_2, and y are position (in m), velocity (in m/s), and head position (in number of tracks), respectively. The discrete-time state-space equations described in (6.7)–(6.8) are obtained by discretizing (6.29)–(6.30) with a Zero-Order Hold (ZOH). (6.3) is used as the feedback controller, and the state-space matrices in (6.9)–(6.10) can be described as

$$A_k = \begin{bmatrix} 1 & 0 & 0 \\ 0 & 0.719 & 0 \\ 0 & 0 & 0.1 \end{bmatrix}, \quad B_k = \begin{bmatrix} 1 \\ 1 \\ 1 \end{bmatrix}, \tag{6.31}$$

$$C_k = \begin{bmatrix} 2.56 \times 10^{-5} & -1.97 \times 10^{-3} & 4.82 \times 10^{-3} \end{bmatrix}, D_k = 3.25 \times 10^{-3} \tag{6.32}$$

It is worth noting that the state-space representation of the feedback controller is not unique. However, the control performance does not depend on the realization [2]. Q_e in (6.25) is given by

$$Q_e = \mathrm{diag}[q_1, q_2, 0, \ldots, 0], \tag{6.33}$$

where q_1 and q_2 are weights for the position x_1 and the velocity x_2, respectively. The elements of Q_e corresponding to the controller state variables are selected to be zero because the responses of the controller state variables are not of concern in this design framework. After some trial and error, the design parameters are chosen as

$$q_1 = 1 \times 10^8, \quad q_2 = 1, \quad q_u = 1, \tag{6.34}$$

and hence we have

$$x_k[0] = \begin{bmatrix} -13.6 & 3.42 & 1.30 \end{bmatrix}^T r. \tag{6.35}$$

The time response of the output y and the control input u of the control system with IVC from (6.35) is shown in Figure 6.7 and is indicated by solid lines. From this figure, it can be seen that the overshoot is drastically reduced while the disturbance response is kept intact. This method is easy to implement because it involves adjustment of the initial values of the feedback controller only.

6.2   Controller Design for Low Acoustic Noise Seek

HDDs are used in various products such as notebook computers, digital audio players, flat-screen televisions, etc. Therefore, there is a growing need for a good control system that moves the R/W heads in HDDs to the target position in a fast and quiet manner. In this section, a short-span seek control system and a long-span seek control system are realized by a new digital control design approach for low acoustic noise.

6.2.1   Short-Span Seek Control for Low Acoustic Noise

Digital control systems are generally used in HDDs and the sampling period for obtaining the head position information cannot be set arbitrarily. As such, the resonant modes above Nyquist frequency are easily excited by fast seeking actions, and these cause acoustic noise and vibration during settling conditions. This chapter introduces a new method that applies the *N-delay control* theory to the feedforward controller design in the TDOF control system [3]. In this method, the control input is changed N times within one sample period [4, 5], and the frequency components of the feedforward input can be adjusted by setting the time widths of the N-delay feedforward input.

The conventional control system shown in Figure 6.8 is considered, and it is a TDOF control system commonly used in track-seeking control in HDDs. The plant $P(s)$ is actuated by feedforward input $u_f[k]$, and the difference between the head position $y_p[k]$ and the reference position $y_{ref}[k]$ brought about by disturbance or model error is suppressed by the feedback controller $C[z]$. These

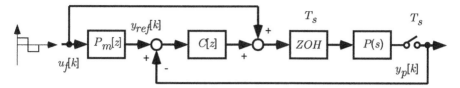

FIGURE 6.8: Conventional TDOF control system.

operations are performed when the position information is obtained and are synchronized with the sample period T_s. In order to realize a high-speed seek operation without exciting the high-order resonant modes of the plant, it is necessary to provide the actuator with a feedforward input $u_f[k]$ that does not contain the frequency components of the resonant modes. The control system shown in Figure 6.9 is considered in this section.

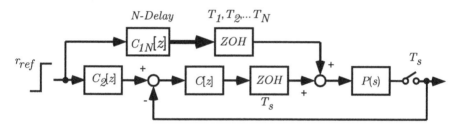

FIGURE 6.9: N-Delay TDOF control system.

In this method, the N-Delay control theory is used for designing the feedforward controller $C_{1N}[z]$ because the feedforward input can be given to the plant regardless of the sample period T_s. The time widths $(T_1 \sim T_N)$ of the N-delay feedforward are determined so that the feedforward input does not contain the frequency components of the resonant modes of the plant $P(s)$. In Figure 6.9, $C_2[z]$ is a filter that calculates the reference position $y_{ref}[k]$ at the sampling period of T_s, and $C[z]$ is a feedback controller calculated at a sample period T_s. As such, the robust performance against disturbances or modeling errors is the same as that of the conventional method. The design objective is thus to design the N-delay controller $C_{1N}[z]$.

The minimal realization of the feedforward controller $C_1(s)$ designed in the continuous-time domain is given by

$$\dot{x}_f = A_f x_f + B_f r_{ref}, \tag{6.36}$$
$$y_f = C_f x_f + D_f r_{ref}, \tag{6.37}$$

and the nominal plant $P_n(s)$ is expressed as

$$\dot{x}_p = A_p x_p + B_p u_f, \tag{6.38}$$
$$y_p = C_p x_p. \tag{6.39}$$

The order of $P_n(s)$ is n_p. (A_p, B_p) is assumed to be controllable, and the order of $C_1(s)$ is n_f. From these equations, the states of the series connection of $C_1(s)$ and $P_n(s)$ at the sample points are given by

$$\begin{bmatrix} x_{pz}[k+1] \\ x_{fz}[k+1] \end{bmatrix} = \begin{bmatrix} e^{A_p T_s} & A_{12z} \\ 0 & e^{A_f T_s} \end{bmatrix} \begin{bmatrix} B_{11z} \\ B_{21z} \end{bmatrix} r_{ref}, \tag{6.40}$$

$$\begin{bmatrix} e^{A_p T_s} & A_{12z} \\ 0 & e^{A_f T_s} \end{bmatrix} = e^{A_{pc} T_s}, \quad A_{pc} = \begin{bmatrix} A_p & B_p C_f \\ 0 & A_f \end{bmatrix}, \tag{6.41}$$

$$\begin{bmatrix} B_{11z} \\ B_{21z} \end{bmatrix} = \int_0^{T_s} e^{A_{pc} \tau} d\tau \begin{bmatrix} B_p D_f \\ B_f \end{bmatrix}, \tag{6.42}$$

where $x_{pz}[k]$ is the state of $P_n(s)$ and $x_{fz}[k]$ is the state of $C_1(s)$ at the sample points. If the feedforward input injected into $P_n(s)$ changes N times within one sample period T_s and N is larger than n_p, the N-delay controller $C_{1N}[z]$ can be written as

$$x_{fzN}[k+1] = e^{A_f T_s} x_{fzN}[k] + B_{fz} r_{ref}, \tag{6.43}$$

$$\begin{bmatrix} u_{fz1}[k] \\ \vdots \\ u_{fzN}[k] \end{bmatrix} = C_{fz} x_{fzN}[k] + D_{fz} r_{ref}, \tag{6.44}$$

where $x_{fzN}[k]$ is the state. B_{fz}, C_{fz}, and D_{fz} are matrices to be determined. Moreover, when the feedforward input from $C_{1N}[z]$ is injected into $P_n(s)$ at N times within one sample period T_s, as shown in Figure 6.10, $P_n(s)$ can be discretized and rewritten as

$$x_{pzN}[k+1] = e^{A_p T_s} x_{pzN}[k] + \underbrace{\begin{bmatrix} H_1 & \cdots & H_N \end{bmatrix}}_{H} \begin{bmatrix} u_{fz1}[k] \\ \vdots \\ u_{fzN}[k] \end{bmatrix}, \tag{6.45}$$

where x_{pzN} is the state of P_n at the sample point when the feedforward input is applied from $C_{1N}[z]$ to $P_n(s)$, and $H_1 \sim H_N$ is decided by setting the time widths $(T_1 \sim T_N)$ of u_{fzN}. H is the matrix of $n_p \times N$ whose elements are given by

$$H_1 = \int_{T_s - T_1}^{T_s} e^{A_p} d\tau B_p, \quad \cdots, \quad H_N = \int_0^{T_s - T_1 - T_2 \cdots - T_N} e^{A_p} d\tau B_p. \tag{6.46}$$

From these equations, the states at the sample point when the output of $C_{1N}[z]$ is given to $P_n(s)$ can be described as

$$\begin{bmatrix} x_{pzN}[k+1] \\ x_{fzN}[k+1] \end{bmatrix} = \begin{bmatrix} e^{A_p T_s} & HC_{fz} \\ 0 & e^{A_f T_s} \end{bmatrix} \begin{bmatrix} x_{pzN}[k] \\ x_{fzN}[k] \end{bmatrix} + \begin{bmatrix} HD_{fz} \\ B_{fz} \end{bmatrix} r_{ref}. \tag{6.47}$$

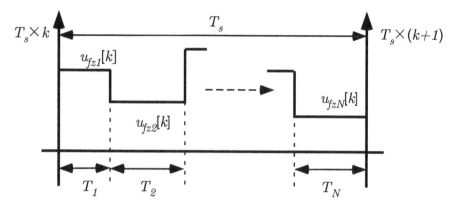

FIGURE 6.10: N-Delay control inputs.

To ensure that $x_{pzN}[k]$ and $x_{fzN}[k]$ agree with $x_{pz}[k]$ and $x_{fz}[k]$ when actuating the nominal model $P_n(s)$ using the feedforward input from N-delay controller $C_{1N}[z]$, the following conditions

$$HC_{fz} = A_{12z}, \quad HD_{fz} = B_{11z}, \quad B_{fz} = B_{21z}, \tag{6.48}$$

are required. Using *the generalized inverse matrix $H^\dagger$ of H*, the solutions of the above equations are given by

$$C_{fz} = H^\dagger A_{12z}, \quad D_{fz} = H^\dagger B_{11z}. \tag{6.49}$$

Note that in order for the generalized inverse matrix $H^\dagger$ to exist,

$$\text{rank}(H) = n_p \tag{6.50}$$

has to be satisfied.

If N is greater than the order n_p and (A_p, B_p) is controllable, (6.50) will definitely hold. From these considerations, the N-delay feedforward controller $C_{1N}[z]$ can be obtained when the time widths ($T_1 \sim T_N$) are decided.

Next, the method for setting the time widths is described. The frequency component of the feedforward input from $C_{1N}[z]$ can be adjusted by using the degree of freedom in setting the time widths ($T_1 \sim T_N$). In order to avoid excitation of the resonant modes in the actuator, it is necessary that the feedforward input does not include the frequency components of the resonant modes of the actuator. We consider the following cost function J given by

$$J = \sum_{\ell=0}^{m} |U_{fz}(j\omega_\ell)|, \tag{6.51}$$

where ω_ℓ is the frequency of the resonant modes of the plant and m is the number of resonant modes to be considered. $U_{fz}(j\omega)$ is the Fourier transform of the feedforward input

$$\{\ldots, u_{fz1}[k], u_{fz2}[k], \ldots, u_{fzN}[k],$$
$$u_{fz1}[k+1], u_{fz2}[k+1], \ldots, u_{fzN}[k+1], \ldots\}. \qquad (6.52)$$

When this cost function is minimized subject to the constraint

$$T_1 + T_2 + \ldots + T_N = T_s, \quad T_i > 0, \qquad (6.53)$$

the time widths $(T_1 \sim T_N)$ that reduce the frequency component at ω_ℓ can be obtained. This problem can be formulated as a constrained nonlinear optimization problem, and can be readily solved using *Sequential Quadratic Programming* (SQP) techniques.

Next, two examples where this method is applied to one track seek are shown. The HDD used for the experiment is 2.5" in diameter, has a track width of 7.2 μm, rotational speed of 4200 rpm, servo sector of 50, and a sampling frequency of 3.5 kHz. By measuring the frequency characteristics of the HDD used for the experiment, the primary bending mode is found to exist at about 4.2 kHz and the secondary bending mode exists at about 8.5 kHz. N is set to three when only the primary bending mode is considered, and N is set to four when the primary and secondary bending modes are considered.

First, the results when only the primary bending mode is considered are shown. The cost function is set as

$$J = |U_{fz}(j \cdot 4000 \cdot 2\pi)|, \qquad (6.54)$$

and the initial time widths are set to

$$T_1 = T_2 = T_3 = \frac{1}{3}T_s. \qquad (6.55)$$

The optimum time widths T_1, T_2, and T_3 are optimized using SQP under the constraint depicted in (6.53) so that the cost function in (6.54) is minimized. The optimum time widths are found to be

$$T_1 = 97.1 \ \mu s, \quad T_2 = 108.5 \ \mu s, \quad T_3 = 80.1 \ \mu s. \qquad (6.56)$$

The contour plot of J is shown in Figure 6.11. In Figure 6.11, the $\circ$ mark and the $\times$ mark indicate the initial value and the optimal solution, respectively. The frequency characteristics of the feedforward input when the optimum solution is applied are shown in Figure 6.12.

The frequency component at 4 kHz is eliminated and low acoustic noise is expected. The results when both the primary and secondary bending modes are considered are shown next. The cost function is set as

$$J = |U_{fz}(j \cdot 4000 \cdot 2\pi)| + |U_{fz}(j \cdot 9000 \cdot 2\pi)|, \qquad (6.57)$$

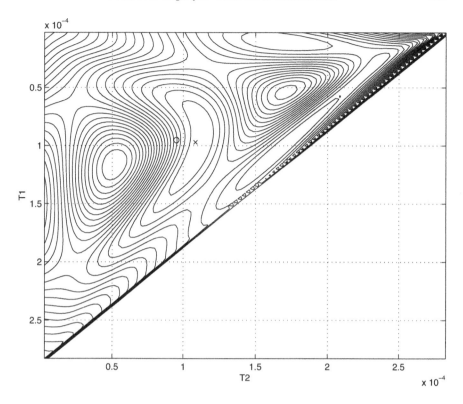

FIGURE 6.11: Contour plot of J.

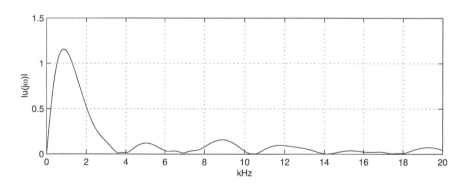

FIGURE 6.12: Frequency characteristics of 3-delay feedforward control input.

and the initial time widths are chosen as

$$T_1 = T_2 = T_3 = T_4 = \frac{1}{4}T_s. \tag{6.58}$$

The optimum time widths T_1, T_2, T_3, and T_4 are found to be

$$T_1 = 68.8 \ \mu s, \ T_2 = 77.4 \ \mu s, \ T_3 = 83.3 \ \mu s, \ T_4 = 56.2 \ \mu s. \qquad (6.59)$$

The frequency characteristics of the feedforward input when the optimum solution is applied are shown in Figure 6.13, and it is confirmed from the figure that multiple frequency components of the feedforward input can be decreased by using the optimum solution.

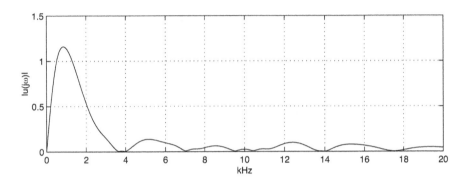

FIGURE 6.13: Frequency characteristics of 4-delay feedforward control input.

In order to evaluate the effectiveness of this approach, several one track-seek experiments are conducted. The head position drawn for one hundred times are shown in Figures 6.14(a)–6.14(c), and the power spectra of the acoustic noise are shown in Figures 6.15(a)–6.15(c).

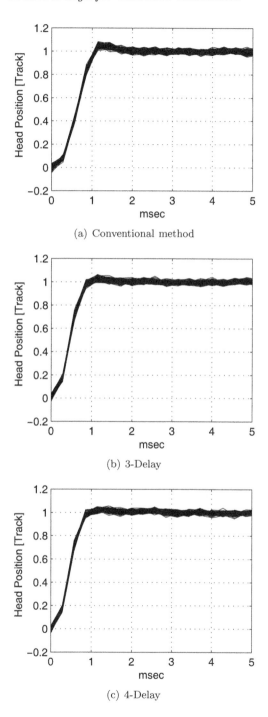

(a) Conventional method

(b) 3-Delay

(c) 4-Delay

FIGURE 6.14: Head positions.

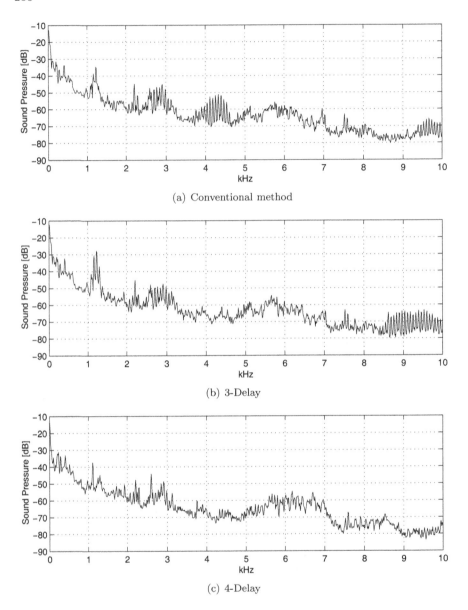

(a) Conventional method

(b) 3-Delay

(c) 4-Delay

FIGURE 6.15: Acoustic noise.

The response of the head position using the N-delay control input is almost the same as that of the conventional method. However, some differences can be observed in the power spectra of the acoustic noise. It is clear that the primary and secondary bending modes are excited when using the conventional method. On the other hand, it is clear that the acoustic noise component at about 4 kHz (which is the primary bending mode of the VCM) is improved by using N-delay control input from Figures 6.15(b) and 6.15(c). In spite of this, Figure 6.15(b) shows that the secondary bending mode is excited in the case of 3-delay control, although the primary bending mode is not excited. The reason for this result is that the 3-delay control input contains many frequency components between 8 kHz and 10 kHz. When the time widths are chosen so that the components of the feedforward input at about 4 kHz and about 9 kHz can be eliminated by using 4-delay control, the components of acoustic noise at about 4 kHz and 9 kHz are decreased. These experimental results show that high-speed seek does not excite the multiple resonant modes above the Nyquist frequency when applying this proposed method to a one track-seek control.

6.2.2 Long-Span Seek Control for Low Acoustic Noise

In general, HDDs have two track-seeking control modes, namely, the *short-span seek control* mode where the seek distance is less than one hundred tracks and the *long-span seek control* mode where the seek distance is more than one hundred tracks. In the short-span seek control mode, the feedforward input and reference position are calculated in advance and stored in the memory of Central Processing Unit (CPU). On the other hand, the feedforward input and reference position should be calculated in real time in the long-span seek mode as the long-span seek mode deals with many seek distances. Therefore, it is very difficult to generate the feedforward input and reference position signals online which do not excite the resonant modes of the plant by using an optimum method that is used in the short-span seek control mode. This section introduces the long-span seek control system where *the frequency-shaped switching line* is applied [6]. The target velocity curve used in the long-span seek control can be regarded as the switching line of the *sliding mode control system* [7]. As such, the frequency-shaped switching line is applied to the long-span seek control to prevent the feedforward input from exciting the resonant modes of the plant.

The basic control system that is applied to the model-following control system is shown in Figure 6.16. In Figure 6.16, $C[z]$ is the feedback controller, f_s is the sampling frequency, $P[z]$ is the discretized nominal model, $P(s)$ is the plant, k is the velocity feedback gain in a model control system that is enclosed by a dotted line, v_{ref} is the target velocity curve, $\uparrow N$ is the up-sampler, and $\downarrow N$ is the down-sampler. The up-sampler is of the ZOH type. The sample period of the model control system is N times that of f_s in order to generate a smooth control input that does not excite the resonant modes of the plant easily, and the input to $P[z]$ is given to $P(s)$ as the feedforward input. The model position is resampled with the sample frequency f_s, and it is injected into the feedback controller as the reference position. During long-span seek control, the current at acceleration is saturated to shorten the seek time. However, the error between the head position and reference position is not suppressed by the feedback controller when the current is saturated,

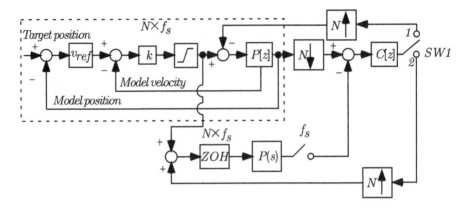

FIGURE 6.16: Model-following control.

and the settling condition is worsened. To keep the head position close to the reference position, the output of the feedback controller is also injected into the model at the early stage of the seek by connecting *SW 1* to 1, and the seek control system is changed to TDOF (by connecting *SW 1* to 2) when the current is not saturated in the later stage of the seek. By applying this method, the influence of the current saturation is reduced and the settling condition is improved. The influence of the resonant modes of the plant should also be considered here since the output of the feedback controller $C[z]$ is given to the model at the early stage of the seek, as shown in Figure 6.16.

The HDD used for the experiment is 2.5" in diameter and has a track width of 0.7 μm with a sampling frequency of 11.52 kHz. The frequency characteristics of the plant and the model are shown in Figure 6.17. The plant has many resonant modes in the high frequency region, and the rigid body mode model of $P[z]$ can be described by a double integrator. The control period of the velocity control in the model control system is half the sampling period, i.e., $N = 2$. Next, we consider the calculation of the frequency characteristics from $u[k]$ to $y_m[k]$ in Figure 6.18.

It is difficult to calculate the frequency characteristics from $u[k]$ to $y_m[k]$ as this system has multiple control periods. As such, it is identified by injecting a sine wave into $u[k]$ and the result is shown in Figure 6.19.

The model in the control system shown in Figure 6.16 is affected by the resonant modes of the plant. Therefore, the design of the velocity control in the model control system should take into consideration the resonant modes of the plant. However, it is very difficult to consider both shortening the seek time and avoiding the excitation of the resonant modes of the plant when designing the target velocity curve. The target velocity curve to the remaining track is calculated in the conventional velocity feedback control. This target velocity curve is the same as the switching line on the phase plane that consists of a position and a velocity in sliding mode control. This means that sliding mode

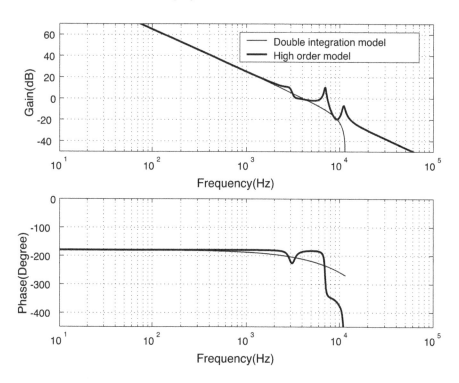

FIGURE 6.17: Frequency responses of VCM plant and model.

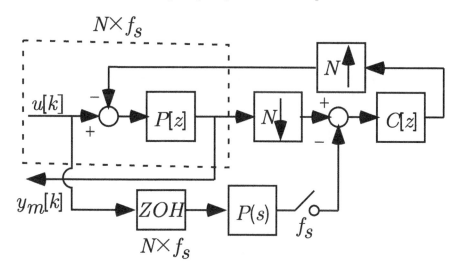

FIGURE 6.18: VCM model in the model control system.

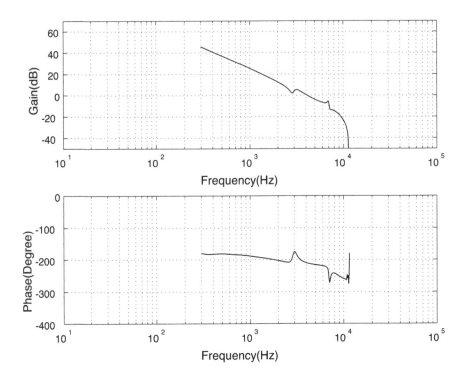

FIGURE 6.19: VCM model used in the model-following control.

control can be easily applied to the long-span seek control. The frequency-shaped switching line is particularly effective in avoiding the excitation of the resonant modes of the plant. The nominal model is used in the design, and the model $P[z]$ is described by a double integrator which can be expressed as

$$\begin{bmatrix} x_1[k+1] \\ x_2[k+2] \end{bmatrix} = \begin{bmatrix} A_{11} & A_{12} \\ A_{21} & A_{22} \end{bmatrix} \begin{bmatrix} x_1[k] \\ x_2[k] \end{bmatrix} + \begin{bmatrix} B_1 \\ B_2 \end{bmatrix} u[k], \qquad (6.60)$$

where $x_1[k]$ is the position and $x_2[k]$ is the velocity. The switching line for this model is set as

$$\varphi[k] = -S(x_1[k]) - x_2[k], \qquad (6.61)$$

where $S(x_1[k])$ has dynamics expressed as

$$z[k+1] = Fz[k] + Gx_1[k], \qquad (6.62)$$
$$S[k] = Hz[k] + Lx_1[k]. \qquad (6.63)$$

From the above equations, the augmented system can be written as

$$
\begin{bmatrix} z[k+1] \\ x_1[k+1] \\ x_2[k+1] \end{bmatrix} = \begin{bmatrix} F & G & 0 \\ 0 & A_{11} & A_{12} \\ 0 & A_{21} & A_{22} \end{bmatrix} \begin{bmatrix} z[k] \\ x_1[k] \\ x_2[k] \end{bmatrix} + \begin{bmatrix} 0 \\ B_1 \\ B_2 \end{bmatrix} u[k], \tag{6.64}
$$

$$
\varphi[k] = -\begin{bmatrix} H & L \end{bmatrix} \begin{bmatrix} z[k] \\ x_1[k] \end{bmatrix} - x_2[k]. \tag{6.65}
$$

In the *sliding condition*,

$$
\varphi[k] = \varphi[k+1] \tag{6.66}
$$

is set up. From (6.64)–(6.66), it can be derived that

$$
Hz[k] + Lx_1[k] + x_2[k] = HFz[k] + (LA_{12} + A_{22})x_2[k]
$$
$$
+(HG + LA_{11} + A_{21})x_1[k] + (LB_1 + B_2)u[k], \tag{6.67}
$$

and the *equivalent linear control input* $u_{eq}[k]$ can be expressed as

$$
u_{eq}[k] = (LB_1 + B_2)^{-1}\{(H - HF)z[k] + (L - HG - LA_{11} - A_{21})x_1[k]
$$
$$
+(1 - LA_{12} - A_{22})x_2[k]\}. \tag{6.68}
$$

Next, the sliding controller is determined. The nonlinear controller to be applied is given by

$$
u_{nl} = \alpha_1 \frac{\varphi[k]}{|\varphi[k]| + \beta} + \alpha_2 \varphi[k], \tag{6.69}
$$

where β is used to prevent chattering in the sliding mode condition and $\alpha_2\varphi[k]$ is used to adjust the time taken to reach the sliding mode condition. When α_2 is larger, the model can arrive at the switching line rapidly, but the transition from acceleration to deceleration becomes steep and the resonant modes of the plant can be easily excited. The long-span seek control system based on the sliding mode control is shown in Figure 6.20.

In Figure 6.20, $v[z]$ defined by (6.62)–(6.63) is the transfer function that gives the frequency characteristics, as shown in Figure 6.21, and k_3, k_4, and k_5 are expressed as

$$
k_3 = (LB_1 + B_2)^{-1}(H - HF), \tag{6.70}
$$
$$
k_4 = (LB_1 + B_2)^{-1}(1 - LA_{12} - A_{22}), \tag{6.71}
$$
$$
k_5 = (LB_1 + B_2)^{-1}(L - HG - LA_{11} - A_{21}). \tag{6.72}
$$

When the initial states of the model are set to $x_1[0] = r$ and $x_2[0] = 0$, the model state is moved from $x_1 = r$ to $x_1 = 0$.

Simulations are carried out to evaluate the effectiveness of this approach. The plant with resonant modes and the model in the velocity control system are the same as that shown in Figure 6.17. The seek distance is one-third of

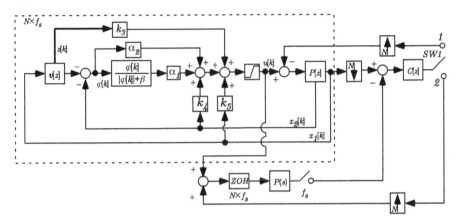

FIGURE 6.20: Multi-rate model-following control system with sliding mode control.

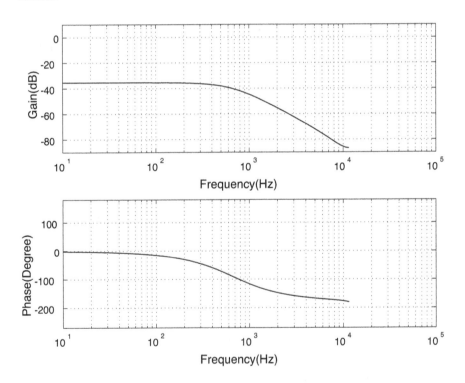

FIGURE 6.21: Frequency characteristics of S.

all tracks, and $S(x_1[k])$ is set to be an LPF, as shown in Figure 6.21. α_1, α_2, and β are tuned while observing the current waveform so that the seek time is the same as that of the conventional method. The current waveform of the proposed method and that of the conventional method are shown in Figures 6.22(a) and 6.22(b), respectively.

The transition from acceleration to deceleration in the conventional method is steeper than that of the proposed method. The results of the Fast Fourier Transform (FFT) analysis of the current are shown in Figures 6.23(a) and 6.23(b).

The high frequency components of the current are removed in the proposed method since the transition from acceleration to deceleration is smooth. The head velocity profiles are shown in Figures 6.24(a) and 6.24(b).

Vibration at settling condition is reduced considerably in the proposed method because the high frequency components of the current are removed. From these simulation results, acoustic noise at the long-span seek mode is expected to be reduced without reduction in seek speed. To confirm the effects of the proposed method, seek experiments are conducted using the same parameters as those used in simulation studies. Figure 6.25 shows the one-third octave analysis of acoustic noise when one-third of all track seek is repeated. A microphone is set up at 1 cm away from the side of the HDD casing, and acoustic noise is recorded through an *A*-weight filter.

It can be seen from Figure 6.25 that acoustic noise over 1 kHz is reduced. This result is almost the same as the results of the FFT analysis in simulations. The overall value is 52.9 dB in the conventional method and 49.8 dB in the proposed method. As such, acoustic noise is reduced by 3 dB. The head positions are shown in Figures 6.26(a) and 6.26(b).

Similarly, these figures are redrawn for one hundred times and the seek time is 12.3 ms for both cases. It is clear that the proposed method can realize low acoustic noise seek without reduction in seek speed.

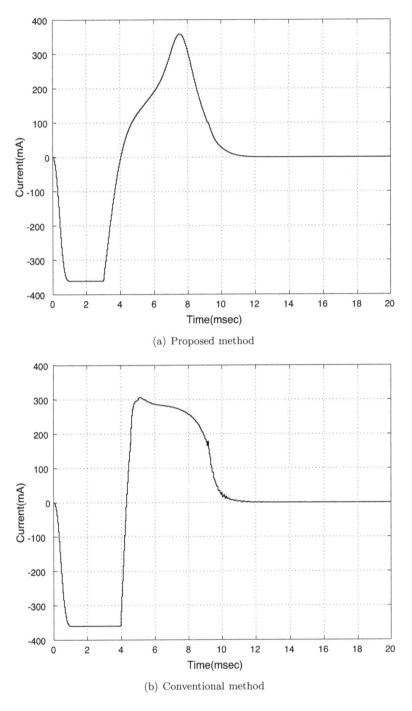

(a) Proposed method

(b) Conventional method

FIGURE 6.22: Current waveforms.

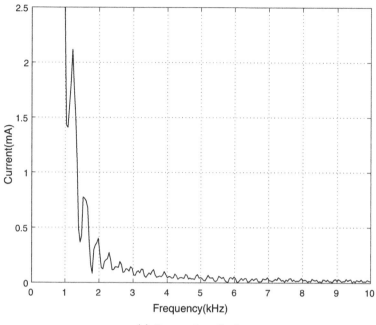

(a) Proposed method

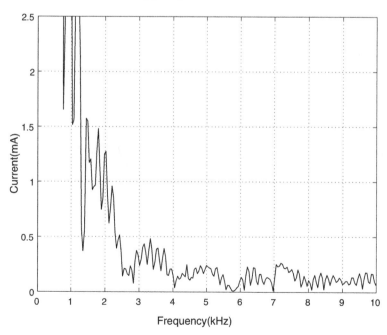

(b) Conventional method

FIGURE 6.23: FFT analysis of currents.

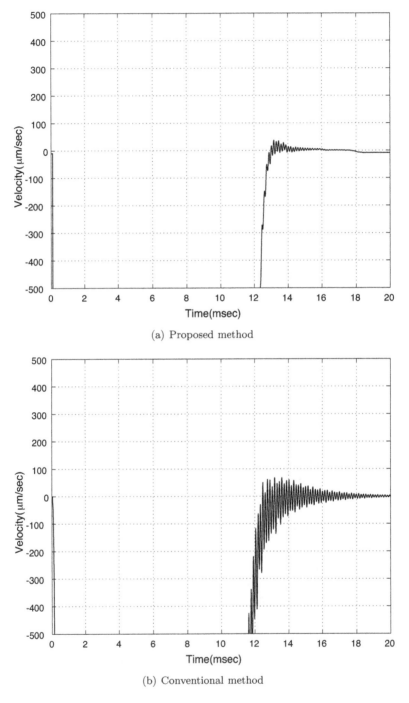

(a) Proposed method

(b) Conventional method

FIGURE 6.24: Velocity profiles.

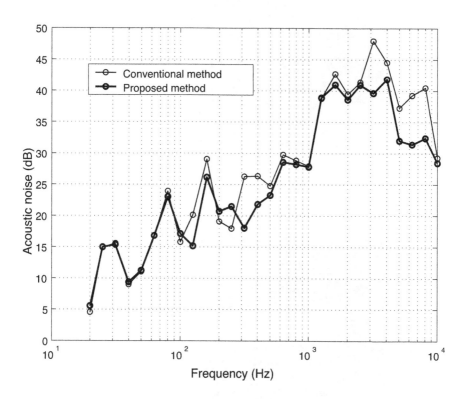

FIGURE 6.25: Acoustic noise.

242

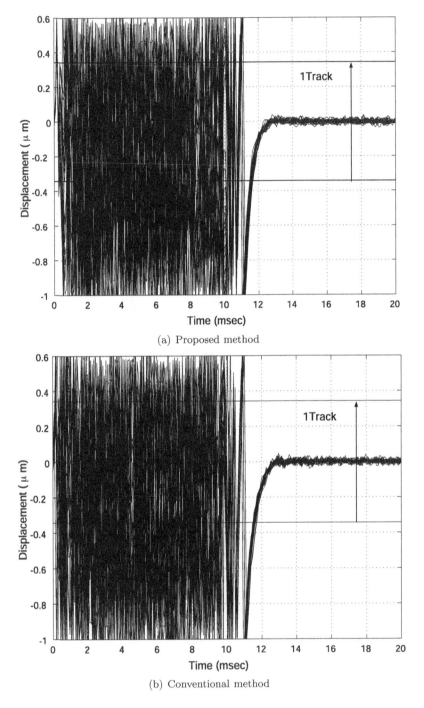

(a) Proposed method

(b) Conventional method

FIGURE 6.26: Head positions.

6.3 Servo Control Design Based on SRS Analysis

In this section, Shock Response Spectrum (SRS) analysis is used in the design of a track-seeking servo controller that minimizes seeking noise in HDDs. The fundamentals of SRS analysis are presented and used to design the reference trajectory for a short-span seeking control system based on Perfect Tracking Control (PTC). The effectiveness of the proposed technique in reducing acoustic noise is verified with experiments on HDDs.

6.3.1 Seeking Noise

Figure 6.27 shows the process where seeking noise is generated. Supplying a current to the VCM generates the driving force which moves the head. When the VCM moves the head, mechanical vibrations occur in the HDD. These vibrations spread to the top cover. It has been concluded that the vibration of the top cover in contact with external air is the main cause of seeking noise.

Seeking noise is detected and evaluated using sound pressure. The relation between the seeking current and sound pressure is equivalent to the relation between the cause (input) and the result (output). In order to reduce seeking noise, it is thus necessary to optimize the seeking current.

Figure 6.28 shows an example of seeking noise, where (a) is a plot of the current and (b) is a plot of the normalized sound pressure. The normalized sound pressure is high in three parts. The duration of the sound pressure generated in each part is short. Therefore, it is considered that seeking noise originates in transient vibrations (not steady-state vibrations) of the mechanical parts.

6.3.2 Concept and Procedure of SRS Analysis

Conventionally, SRS analysis has been used for damage-boundary testing of products, seismic wave analysis, etc. It is a method for predicting the response of a structure subjected to a transient signal [8, 9, 10, 11]. The concept of SRS analysis is shown in Figure 6.29. In SRS analysis, it is assumed that a real structure consists of many One-Degree-of-Freedom (ODOF) systems. Each ODOF system has a different resonant frequency. The procedure for SRS analysis consists of the following three steps:

1. Time history responses of each ODOF system to a given transient input signal are calculated;

2. Various peak levels of the time-history responses are measured; and

3. Peak levels are plotted against the resonant frequencies.

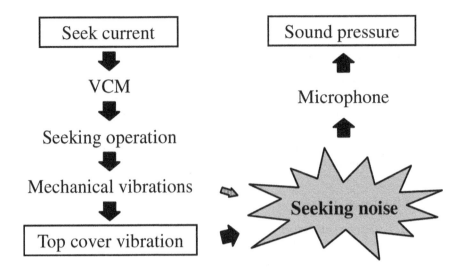

FIGURE 6.27: Seeking noise generating process.

The peak level over all time (which refers to both peak levels during and after the transient signal) is known as the Initial Shock Spectrum (ISS). The peak level after the transient signal termination is known as the Residual Shock Spectrum (RSS).

By carrying out SRS analysis of the seeking current, the resonant frequencies that cause seeking noise can be predicted. Since the mechanical vibrations during seeking cause seeking noise, the resonant frequencies that cause seeking noise can be predicted from the ISS. Similarly, since the mechanical vibrations after seeking cause residual vibrations, the resonant frequencies that cause residual vibrations can be predicted from the RSS.

6.3.3 Models for SRS Analysis

The absolute acceleration model generally used for conventional SRS analysis is given by

$$F_{aai}(s) = \frac{2\zeta(2\pi \times f_i)s + (2\pi \times f_i)^2}{s^2 + 2\zeta(2\pi \times f_i)s + (2\pi \times f_i)^2},$$ (6.73)

where f_i represents the ith resonant frequency (logarithmically spaced at integer fractions of an octave) and ζ is the damping ratio (normally $\zeta = 5\%$). The same damping value is used for each ODOF system. (6.73) is the transfer function from the base acceleration to the absolute acceleration of each ODOF system. For the case where SRS analysis is used for damage-boundary testing of products, it is appropriate to evaluate the absolute acceleration that is added to each ODOF system. As such, the *absolute acceleration model*

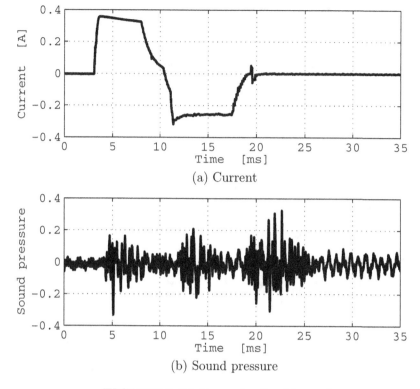

(a) Current

(b) Sound pressure

FIGURE 6.28: Example of seeking noise.

has been used for the SRS analysis. However, it is preferable to compare the responses of each ODOF system relatively when SRS analysis is used for predicting seeking noise. In order to satisfy this purpose, it is appropriate to use the following *relative acceleration model*

$$F_{rai}(s) = \frac{s^2}{s^2 + 2\zeta(2\pi \times f_i)s + (2\pi \times f_i)^2}. \tag{6.74}$$

(6.74) is the transfer function from the base acceleration to the relative acceleration of each ODOF system.

Figure 6.30 shows an example of the frequency characteristics of both the absolute acceleration model and the relative acceleration model, where the resonant frequency is 1 kHz and the damping ratio is 5%. The absolute acceleration model has a *low-pass* filter characteristic, while the relative acceleration model has a *high-pass* filter characteristic. Therefore, the relative acceleration model can eliminate the influence of the DC component from the seeking current and can relatively evaluate the responses of each ODOF system. Hereafter, the relative acceleration model is used for SRS analysis of the seeking current.

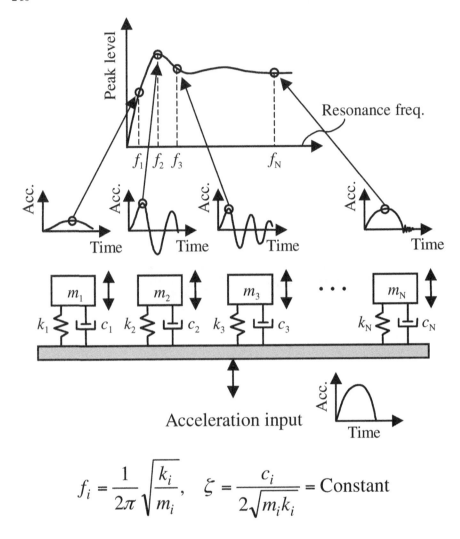

$$f_i = \frac{1}{2\pi}\sqrt{\frac{k_i}{m_i}}, \quad \zeta = \frac{c_i}{2\sqrt{m_i k_i}} = \text{Constant}$$

FIGURE 6.29: Concept of SRS analysis.

6.3.4 Examples of SRS Analyses

Examples of SRS analyses are shown in Figure 6.31 where (a) is a plot of the modeled current waveforms and (b) is a plot of SRS analyses of these waveforms. T_d represents a seeking time and f_d is equal to $\frac{1}{T_d}$. The evaluated waveforms are the sine wave, triangular wave, square wave, and low-pass filtered square wave. The amplitude of these waveforms without a Low-Pass Filter (LPF) is adjusted so that the integrated value of the waveform from 0 to $\frac{T_d}{2}$ is identical.

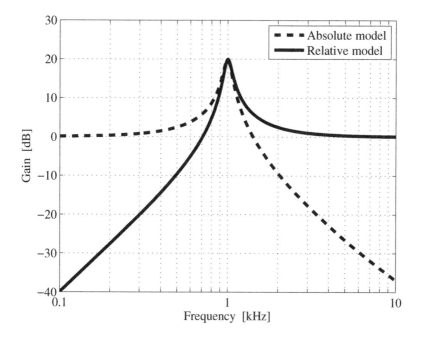

FIGURE 6.30: Frequency response of models for SRS analysis.

The square wave is suitable for reducing the maximum peak level. It is considered that the maximum peak level depends on the maximum amplitude of the waveform. The sine wave is suitable for reducing the peak levels of high resonant frequencies. It is considered that the peak levels of high resonant frequencies depend on the smoothness of the waveform. As such, the smooth waveform limited amplitude is suitable as the current waveform for reducing seeking noise.

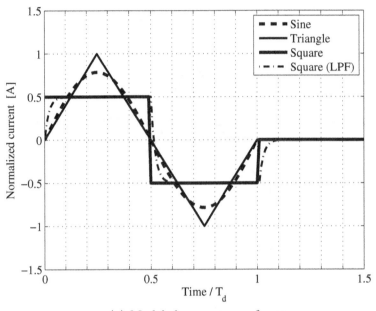

(a) Modeled current waveforms

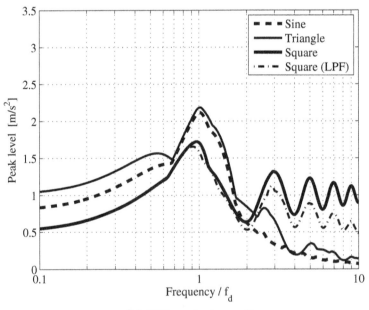

(b) SRS analysis results

FIGURE 6.31: Examples of SRS analysis.

6.3.5 Acoustic Noise Reduction Based on SRS Analysis

The acoustic noise reduction process is shown in Figure 6.32. The seeking current and the sound pressure are measured. The SRS analysis of the seeking current predicts the resonant frequencies. FFT analysis of the sound pressure is used to find the acoustic noise frequencies. The predicted resonant frequencies are compared with the acoustic noise frequencies. When the predicted resonant frequencies are close to the acoustic noise frequencies, it is necessary to optimize the seeking current for reducing the seeking noise. The optimization of the seeking current uses the peak levels as indices, and the seeking control system is re-designed so that the peak levels are reduced.

When the resonant frequencies of the mechanical parts are known *a priori*, they can be used instead of acoustic noise frequencies which are measured from experiments. A loud noise may occur if the mechanical parts of the HDD have the predicted resonant frequencies.

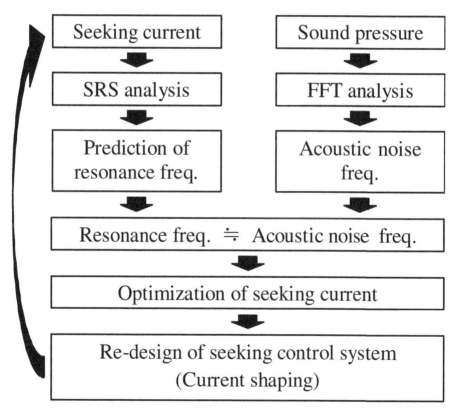

FIGURE 6.32: Acoustic noise reduction based on SRS analysis.

An example of the acoustic noise reduction process is shown below. SRS analysis optimizes the reference trajectory, which is then applied to the short-span seeking control system based on PTC [12].

Figure 6.33 shows the block diagram of a short-span seeking control system based on PTC [11] where T_s is the standard sampling time, u_{fb} is the feedback control input, u_{ff} is the feedforward control input that corresponds to the acceleration, u is the control input, x_r is the reference position trajectory, v_r is the reference velocity trajectory, $P(s)$ is the plant, and $C(z)$ is the feedback controller. The plant model shown in Figure 6.33 is the discrete-time plant model, which includes the multi-rate hold and the delay in both the plant and computation.

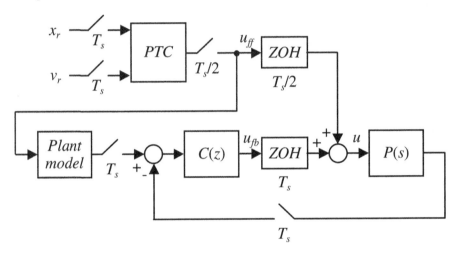

FIGURE 6.33: Block diagram of short-span seeking control system based on PTC.

Figure 6.34 shows the trapezoid acceleration trajectory $a_r(t)$ where

$$\text{Region I}: \qquad a_r(t) = \frac{a_{max}}{T_1} t, \qquad (6.75)$$

$$\text{Region II}: \qquad a_r(t) = a_{max}, \qquad (6.76)$$

$$\text{Region III}: \qquad a_r(t) = \frac{a_{min} - a_{max}}{T_3 - T_2}(t - T_2) + a_{max}, \qquad (6.77)$$

$$\text{Region IV}: \qquad a_r(t) = a_{min}, \qquad (6.78)$$

$$\text{Region V}: \quad a_r(t) = (A_0 + a_{min})\exp\left(-\frac{t - T_4}{ctime}\right) - A_0. \qquad (6.79)$$

$T_1 - T_4$ are the parameters for determining the shape of the trajectory, T_5 is the parameter for determining the seeking time, $ctime$ is the time constant of the exponential function, a_{max} is the maximum acceleration, a_{min} is the minimum acceleration, and A_0 is the adjustment parameter. Both the reference position trajectory and the reference velocity trajectory (which are applied to the short-span seeking control system based on PTC) are obtained by integrating the above trapezoid acceleration trajectory.

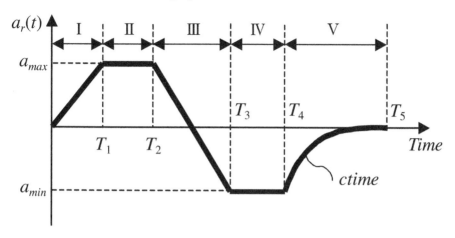

FIGURE 6.34: Trapezoid acceleration trajectory.

When the boundary condition of the trapezoid acceleration trajectory is set to $a_r(T_5) = 0$, the adjustment parameter A_0 becomes

$$A_0 = a_{min} \frac{\exp\left(-\frac{T_5 - T_4}{ctime}\right)}{1 - \exp\left(-\frac{T_5 - T_4}{ctime}\right)}. \tag{6.80}$$

Both the maximum acceleration a_{max} and the minimum acceleration a_{min} are determined according to the following boundary conditions

$$x_r(0) = 0, \quad v_r(0) = 0, \quad x_r(T_5) = L, \text{ and } v_r(T_5) = 0,$$

where L is the seeking span. Therefore, the parameters of the trapezoid acceleration trajectory become $T_1 - T_5$ and $ctime$. By using SRS analysis, $T_1 - T_4$ and $ctime$ are optimized. Since T_5 is the parameter for determining the seeking time, T_5 is fixed. Furthermore, these parameters have the following constraint

$$0 < T_1 \le T_2 < T_3 \le T_4 < T_5. \tag{6.81}$$

This constraint is essential so that the shape of the trajectory becomes a trapezoid.

By carrying out SRS analysis of the seeking current, both RSS and ISS are determined. In order to reduce both the residual vibrations and acoustic noise, the performance index is set as

$$J = \sum_m Q_m \cdot S_{RSS}(f_m) + \sum_n W_n \cdot S_{ISS}(f_n), \tag{6.82}$$

where S_{RSS} is the value of RSS at resonant frequency f_m, S_{ISS} is the value of ISS at resonant frequency f_n, and Q_m and W_n are the weighting values. The resonant frequencies to be considered are set to f_m and f_n. By minimizing the performance index in (6.82), the reference trajectory that reduces both residual vibration and acoustic noise can be determined.

The initial parameters of the trapezoid acceleration trajectory before optimization are determined as

$$T_1 = 1.5T_s, \ T_2 = 3.5T_s, \ T_3 = 7.5T_s, \ T_4 = 8T_s, \ T_5 = 13T_s, \ ctime = 1 \times 10^{-4},$$

and the seeking span is forty tracks. The reference trajectory based on this trapezoid acceleration trajectory is applied to the short-span seeking control system based on PTC, and the seeking current is calculated by carrying out the simulation. Figure 6.35 shows the results of SRS analysis of the seeking current. The resonant frequencies with large peak levels are about 1 kHz.

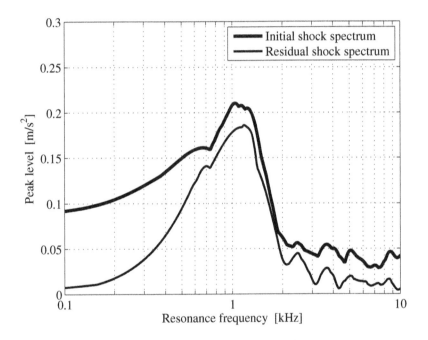

FIGURE 6.35: SRS analysis before optimization.

If the mechanical parts of the plant have resonances at these frequencies, residual vibrations and acoustic noise may occur. Figure 6.36 shows the experimental results before optimization where (a) is a plot of the current, (b) is a plot of the average and envelope of Position Error Signal (PES) in 200-repetition seeking, (c) is a plot of the FFT analysis of the PES, (d) is a plot of the Sound Pressure (SP), and (e) is a plot of the FFT analysis of the SP. The PES is normalized by a track, and the SP is normalized so that its maximum value is one. Both residual vibration and acoustic noise occur in the experiment. The frequencies of the residual vibrations and acoustic noise are identified to be about 880 Hz and 1.3 kHz, respectively.

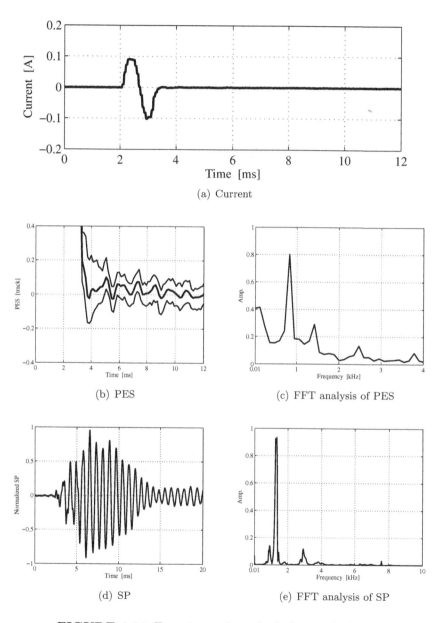

(a) Current

(b) PES

(c) FFT analysis of PES

(d) SP

(e) FFT analysis of SP

FIGURE 6.36: Experimental results before optimization.

In order to reduce both residual vibrations and acoustic noise, the performance index is set as

$$J = S_{RSS}(880\text{Hz}) + S_{ISS}(1.3\text{kHz}), \qquad (6.83)$$

where both weighting values are set as $Q_m = W_n = 1$. To simplify the optimization, $T_1 - T_4$ are set as the integer time of $\frac{T_s}{2}$ and *ctime* is fixed.

The parameters of the trapezoid acceleration trajectory after optimization are determined as

$$T_1 = 0.5T_s, \ T_2 = 1.5T_s, \ T_3 = 8.5T_s, \ T_4 = 9T_s, \ T_5 = 13T_s, \ ctime = 1 \times 10^{-4}.$$

Figure 6.37 shows the results of SRS analysis of the seeking current after optimization. The values of ISS at 2 kHz or less are reduced, and the values of RSS from 500 Hz to 2 kHz are reduced.

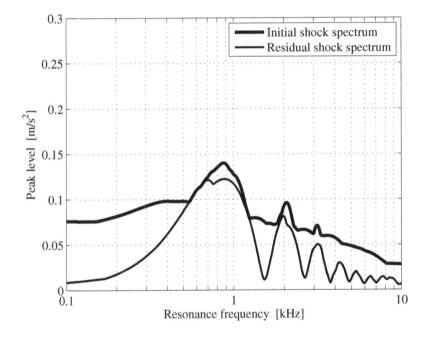

FIGURE 6.37: SRS analysis after optimization.

Figure 6.38 shows the experimental results after optimization. Both residual vibrations at about 880 Hz and acoustic noise at about 1.3 kHz are reduced. These results indicate that the SRS analysis based optimization method can effectively reduce both residual vibration and acoustic noise.

Figure 6.39 shows the effect of different weighting values. In (6.83), the weighting value Q_m of the RSS at 880 Hz is fixed at 1, and the weighting

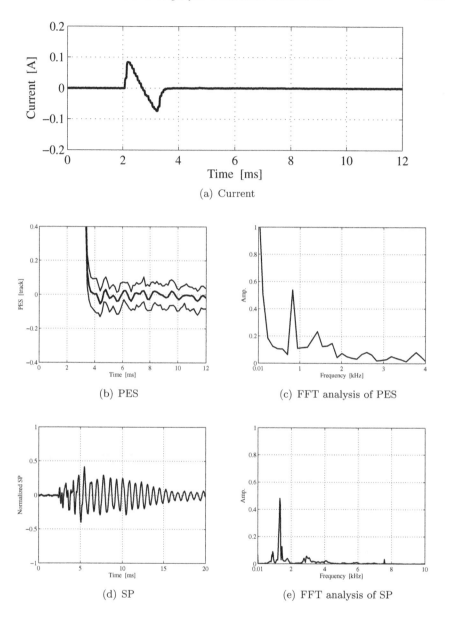

(a) Current

(b) PES

(c) FFT analysis of PES

(d) SP

(e) FFT analysis of SP

FIGURE 6.38: Experimental results after optimization.

value W_n of the ISS at 1.3 kHz is changed. The optimized trajectory is determined according to SRS analysis, and the seeking current is calculated by simulations of the optimized trajectory. By carrying out SRS analysis of the seeking current, the value of the RSS at 880 Hz and the value of the ISS at 1.3

kHz are calculated. These values are plotted in Figure 6.39. When the weighting value W_n becomes large, the value of the ISS at 1.3 kHz becomes small. The effects of different weighting values are verified. However, it seems that there is a trade off between the value of the ISS at 1.3 kHz and the value of the RSS at 880 Hz, i.e., when the weighting value W_n becomes large, the value of the RSS at 880 Hz becomes large. In the case that the main purpose of the design is reduction of residual vibrations, it is appropriate that the weighting value Q_m of the RSS becomes large. On the hand, for the case where the main purpose of the design is acoustic noise reduction, it is appropriate that the weighting value W_n of the ISS becomes large.

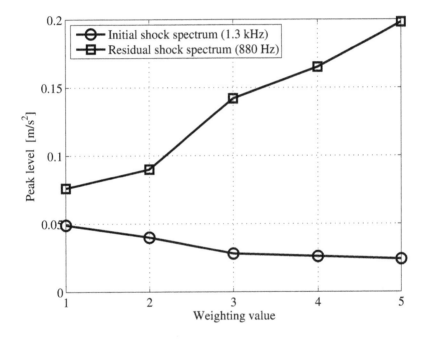

FIGURE 6.39: Effects of different weighting values.

Bibliography

[1] S. C. Wu and M. Tomizuka, "Multi-Rate Digital Control with Interlacing and Its Application to Hard Disk Drive Servo," in *Proceedings of the American Control Conference*, pp. 4347–4352, 2003.

[2] M. Hirata and M. Tomizuka, "Short Track Seeking of Hard Disk Drives under Multirate Control–Computationally Efficient Approach Based on Initial Value Compensation," *IEEE/ASME Transactions on Mechatronics*, Vol. 10, No. 5, pp. 535–545, 2005.

[3] S. Takakura, "Design of the Tracking System using N-Delay Two-Degrees-of-Freedom Control and its Application to Hard Disk Drives," *IEEJ Transactions on Industry Applications*, Vol. 119, No. 5, pp. 728–734, 1999 (in Japanese).

[4] T. Mita and Y. Chida, "MIMO Two-Delay Digital Control and Application," *The Society of Instrument and Control Engineers*, Vol. 24, No. 5, pp. 467–474, 1988 (in Japanese).

[5] H. Fujimoto and A. Kawamura, "New Digital Redesign Method in Use of N-Delay Control," *Transactions of the IEE Japan*, Vol. 117, No. 5, pp. 645–654, 1997 (in Japanese).

[6] S. Takakura and M. Yatsu, "Design of a Low Vibration and Low Acoustic Noise Seek Control Method for Hard Disk Drives," *IEEJ Transactions on Industry Applications*, Vol. 125, No. 12, pp. 1076–1083, 2005 (in Japanese).

[7] K. Nonami and H. Tian, *Sliding Mode Control Theory*, Corona Publishing Co. Ltd., 1994 (in Japanese).

[8] D. O. Smallwood, "An Improved Recursive Formula For Calculating Shock Response Spectra," *The Shock and Vibration Bulletin*, Vol. 51, No. 2, pp. 211–217, 1981.

[9] S. Smith and W. Hollowell, "A Proposed Method to Standardize Shock Response Spectrum (SRS) Analysis," *Journal of the Institute of Environmental Sciences*, Vol. 39, No. 3, pp. 19–24, May-June 1996.

[10] A. Okuyama, T. Hamaguchi, K. Watanabe, T. Horiguchi, K. Shishida, and Y. Nishimura, "Shock-Response-Spectrum Analysis for Acoustic Noise of a Seeking Control System in a Hard Disk Drive," in *Proceedings of The Sixth International Conference on Motion and Vibration Control (MOVIC2002)*, Vol. 1, pp. 320–325, 2002.

[11] A. Okuyama, M. Kobayashi, T. Horiguchi, and K. Shishida, "Reference Trajectory Design Based on Shock Response Spectrum Analysis for a Seeking Control System in Magnetic Disk Drives," *Transactions of the IEE Japan*, Vol. 124-D, No. 1, pp. 116–123, 2000 (in Japanese).

[12] H. Fujimoto and A. Kawamura, "Perfect Tracking Digital Motion Control Based on Two-Degrees-of-Freedom Multirate Feedforward Control," in *Proceedings of the IEEE International Workshop on Advanced Motion Control*, pp. 322–327, 1998.

Chapter 7

HDD Benchmark Problem

Mitsuo Hirata

Utsunomiya University

7.1   Public Release of the HDD Benchmark Problem

The demands for higher positioning accuracy and faster access speed continue to increase in Hard Disk Drive (HDD) control systems. To respond to these demands, the advantages and disadvantages of the conventional methods need to be analyzed and understood to be feedback for the development of advanced control technologies. However, there exist many kinds of HDDs, ranging from low-end HDDs in mobile applications to high-end HDDs for use in enterprise servers. The plant characteristics in these HDDs differ from one another, and the required control performances are also vastly different. As such, it is unfair and difficult to compare the potential of the control methods developed for different types of HDDs.

Furthermore, the HDD industry expects academic researchers to be actively involved in research on advanced control methodologies for HDDs. The plant and disturbance models of actual HDDs are difficult to obtain in reality, without making a research collaboration contract with the HDD companies *a priori*. Even when the required information is obtained, control theorists may find it difficult to embark on research of HDD control as knowledge and know-how on HDDs are required.

With these considerations in mind, the comparisons among various conventional control methods will be fair by using a common plant and disturbance model, if a reference model reflecting various aspects and features of actual HDDs can be constructed. By making this reference model available to the public as a "Benchmark Problem," other control engineers and theorists will also be able to start research on HDD control easily. This will contribute to the growth of the field of HDD control as a result.

To realize these purposes, the "Investigation R&D Committee for Next Generation Servo Control Technology For Mass-Storage System," the Chair being Dr. Takashi Yamaguchi, at the Institute of Electrical Engineers of Japan (IEEJ) set up a working group to brainstorm and develop the HDD Benchmark Problem together. The working group released the HDD Benchmark Problem for the track-following control problem in the **HDD Benchmark Problem Version 1**. The members of the working group consist of researchers and engineers both from academia and industry. All the members from industry, e.g., Hitachi Ltd., Hitachi Global Storage Technologies (HGST), Toshiba Corporation, and Fujitsu Ltd., etc., are working on control systems design for actual HDD products. This allows very precise models of the plant and the disturbances to be defined, reflecting all the physical properties of actual HDDs[1].

In September 2005, a special session of the technical meeting on Industrial Instrumentation and Control in IEEJ was held, and papers on *track-following controller* design were presented [1, 2, 3, 4, 5]. In the following year, the plant model was updated to reflect more realistic properties of actual HDDs, and the **HDD Benchmark Problem Version 2** was released. In September 2006, another special session of the technical meeting on Industrial Instrumentation and Control in IEEJ was held, and papers on short *track-seeking controller* design were presented [6, 7, 8, 9, 10]. Finally in 2010, the plant models in Versions 1 and 2 were unified in the latest version of the **HDD Benchmark Problem Version 3**, and both models can now be used by selecting the corresponding parameter files.

The corresponding MATLAB and Simulink source codes for all three versions of the HDD Benchmark Problem can be downloaded online at http://mizugaki.iis.u-tokyo.ac.jp/nss/MSS_bench_e.htm. In this section, the details of the HDD Benchmark Problem Version 3 are described along with examples of track-following and track-seeking controller design.

---

[1] All the members of the working group wrote this book.

7.2 Plant Model

A 3.5" HDD used in a desktop computer with 100 kTPI is assumed as the plant model, and the block diagram of the plant is shown in Figure 7.1. This model is constructed by assuming that the actuator of the HDD moves in a straight line, but the Voice Coil Motor (VCM) rotates around the pivot bearing in reality. The control input $u(t)$ is a current input to the VCM in A (Amperes), and the head position is the measurement output $y(t)$ in number of tracks. It is assumed that a current amplifier exists between the reference and the actual VCM current.

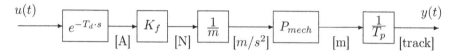

FIGURE 7.1: Block diagram of the HDD plant model.

The values of the physical parameters in Figure 7.1 are shown in Table 7.1. In the HDD Benchmark Problem Version 2, the force constant is rounded off to unity.

TABLE 7.1: Plant Parameters

	Description	Value (Version 1)	Value (Version 2)	Units
T_d	Input delay	1.0×10^{-5}	1.0×10^{-5}	s
K_f	Force constant	9.512×10^{-1}	1.0	N/A
m	Equivalent mass	1.0×10^{-3}	1.0×10^{-3}	kg
T_p	Track pitch	2.54×10^{-7}	2.54×10^{-7}	m

In this case, the transfer function from $u(t)$ to $y(t)$ is given by

$$P_f(s) = K_p P_{mech}(s) e^{-T_d s}, \qquad (7.1)$$

where $K_p = \frac{K_f}{m T_p}$.

The mechanical model $P_{mech}(s)$ of the head actuator in HDDs is defined as

$$P_{mech}(s) = \sum_{i=1}^{N} \frac{A_i}{s^2 + 2\zeta_i \omega_i s + \omega_i^2}, \qquad (7.2)$$

where $w_i = 2\pi f_i$ in rad/s, the input is acceleration in m/s², and the output is position in m. The parameters of (7.2) in the HDD Benchmark Problem Versions 1 and 2 are quite different, and are shown in Tables 7.2 and 7.3, respectively. The mechanical resonant frequencies in Version 1 are defined as

TABLE 7.2: Parameters of P_{mech} in Version 1

i	f_i (Hz)	ζ_i	A_i
1	90	0.5	1.0
2	4100	0.02	−1.0
3	8200	0.02	1.0
4	12300	0.02	−1.0
5	16400	0.02	1.0
6	3000	0.005	0.01
7	5000	0.001	0.03

TABLE 7.3: Parameters of P_{mech} in Version 2

i	f_i (Hz)	ζ_i	A_i
1	90	0.5	1.0
2	3000	0.01	−0.01
3	4100	0.03	−1.0
4	5000	0.01	0.3
5	7000	0.01	−1.0
6	12300	0.005	1.0
7	16400	0.005	−1.0

TABLE 7.4: Upper and Lower Bounds of Plant Parameters in Version 1

f_i	Max (%)	Min (%)
$f_{2,3}$	−15%	+15%
$f_{4,5}$	−10%	+10%
$f_{6,7}$	−5%	+5%
$A_{6,7}$	−200%	+0%

TABLE 7.5: Upper and Lower Bounds of Plant Parameters in Version 2

f_i	Max (%)	Min (%)
f_2	−2%	+1%
f_3	−5%	+2.5%
$f_{4,5,6,7}$	−2%	+0.5%
$A_{2,4}$	−200%	+0%
A_3	−0%	+20%
$\zeta_{5,6,7}$	−50%	+100%

integral multiples of the primary resonant mode, and those in Version 2 are refined to represent actual HDDs as much as possible. The Bode plots of $P_f(s)$ are shown in Figures 7.2 and 7.3.

Parametric variations in the plant are also defined in the HDD Benchmark Problem. The variations of the resonant frequencies f_i, damping ratios ζ_i, and gains A_i are shown in Table 7.4 for Version 1 and Table 7.5 for Version

2. By combining these parametric variations, ten and eighteen possible combinations of variation models are generated by the HDD Benchmark Problem for Versions 1 and 2, respectively. It should be noted that a nominal model is also included in these variation models. The Bode plots of these variation models are shown in Figures 7.4 and 7.5. A change in $\pm10\%$ of loop gain is also defined in the HDD Benchmark Problem.

The plant parameters in Version 1 are suitable for the *track-following problem* as the parametric variations of the resonant modes are exaggerated, which makes robust stabilization challenging and difficult. On the other hand, the plant parameters of Version 2 are suitable for the *track-seeking problem* as the mechanical model is defined such that the residual vibrations after track-seeking mimic those of actual HDDs.

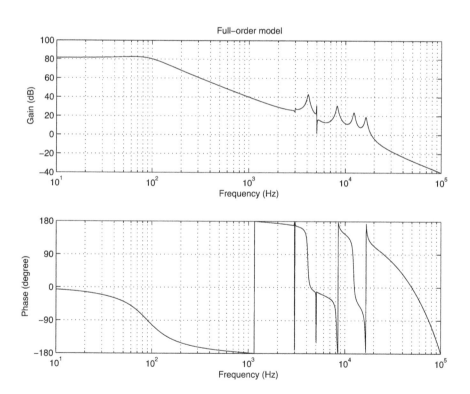

FIGURE 7.2: Frequency response of the nominal model in Version 1.

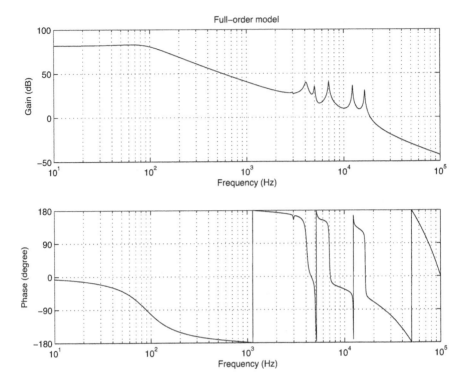

FIGURE 7.3: Frequency response of the nominal model in Version 2.

7.3 Disturbance Model

In this section, the various disturbance sources used in the HDD Benchmark Problem will be introduced and explained in detail. The disturbances defined in the HDD Benchmark Problems are:

1. force disturbance $d_f(t)$;

2. flutter disturbance $d_{flt}(t)$;

3. Repeatable Run-Rout (RRO) $d_{rro}[k]$; and

4. measurement noise $d_n[k]$.

The summing points of these disturbances into the HDD servo system are shown in Figure 7.6. In this figure, $\mathcal{H}_{T_u}$ is a hold with a sampling period of T_u, and $\mathcal{S}_{T_y}$ is a sampler with a sampling period of T_y. The same parameters for disturbances are used for both Versions 1 and 2.

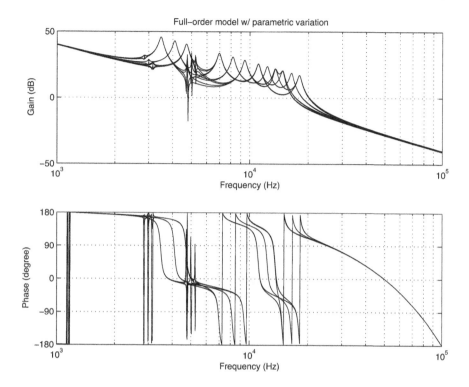

FIGURE 7.4: Frequency responses of perturbed plants with the nominal model in Version 1.

7.3.1 Force Disturbance

The force disturbance $d_f(t)$ is defined as a white Gaussian noise with a standard deviation of $\sigma_{dt} = 1.0 \times 10^{-4}$ A and is injected at the plant input. The time response and spectrum of $y[k]$ when $d_f(t)$ is applied are shown in Figures 7.7 and 7.8, respectively. It should be noted that the frequency response of the plant is reflected in the spectrum because $y[k]$ is observed at the plant output.

7.3.2 Flutter Disturbance

The flutter disturbance $d_{flt}(t)$ is attributed to the disk flutter which is cased by airflow as a result of high-speed disk rotations. This phenomenon causes a relative position error to occur between head and disk. As such, the flutter disturbance can be defined as a disturbance at the plant *output*. In the HDD Benchmark Problem, elastic vibrations of the head, arm, and suspension are also considered as the flutter disturbance.

In the HDD Benchmark Problem, $d_{flt}(t)$ is defined as the output when a

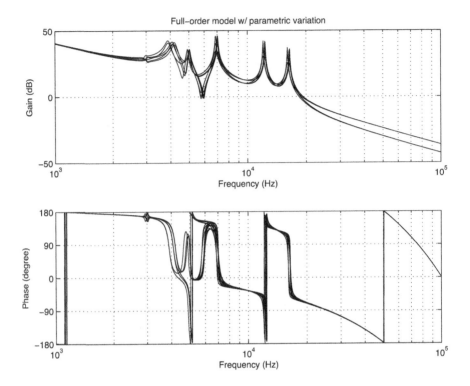

FIGURE 7.5: Frequency responses of perturbed plants with the nominal model in Version 2.

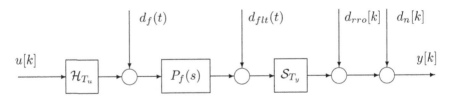

FIGURE 7.6: Disturbances and their summing points.

white Gaussian noise with a standard deviation of $\sigma_{dflt} = 1$ is injected into the transfer function given by

$$G_{flt}(s) = \sum_{i=1}^{N_{flt}} \frac{\alpha_{flt} \, 2 \, \zeta_{flt}^i \, (\omega_{flt}^i)^2}{s^2 + 2\zeta_{flt}^i \omega_{flt}^i s + (\omega_{flt}^i)^2}, \tag{7.3}$$

where $\omega_{flt}^i = 2\pi f_{flt}^i$. The parameters of (7.3) are shown in Table 7.6. The time response and spectrum of $y[k]$ when $d_{flt}(t)$ is applied are shown in Figures 7.7 and 7.8, respectively.

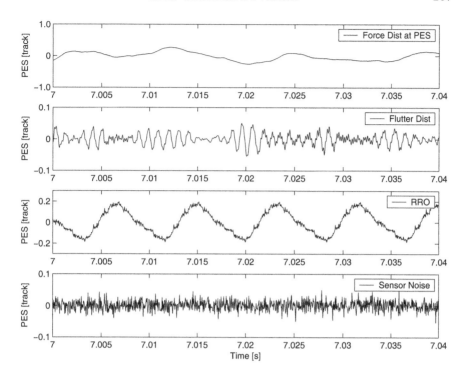

FIGURE 7.7: Time response of $y[k]$ with disturbances and sensor noise.

TABLE 7.6: Flutter Disturbance

i	1	2	3	4	5	6	7	8	9
Frequency (f_{flt}^i)/Hz	750	780	900	1020	1080	1230	1800	3000	5000
Damping ratio (ζ_{flt}^i)	0.01	0.01	0.01	0.01	0.005	0.01	0.005	0.002	0.002
Gain (α_{flt}^i)	0.09	0.17	0.20	0.17	0.06	0.09	0.06	0.06	0.12

7.3.3 RRO

RRO $d_{rro}[k]$ is a position disturbance which is synchronized with the disk rotation and is defined as a discrete-time signal with a sampling period of T_y. For a sector number n_s,

$$d_{rro}[k] = d_{rro}[k + n_s] \qquad (7.4)$$

is satisfied. In the HDD Benchmark Problem, it is assumed that RRO arises as a result of both disk eccentricity and mis-registration of the servo signal as

$$d_{rro}[k] = \sum_{i=1}^{3} \alpha_{rro}^i \sin(2\pi f_{rro}^i k T_s) + \hat{d}_{rro}[k], \qquad (7.5)$$

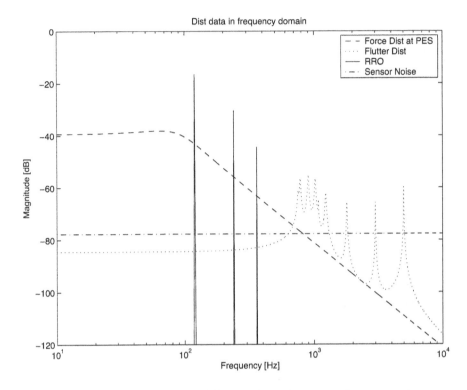

FIGURE 7.8: Spectra of $y[k]$ with disturbances and sensor noise.

where $\hat{d}_{rro}[k]$ corresponds to the mis-registration of the servo signal. $\hat{d}_{rro}[k]$ is a repetitive time-series data where $\hat{d}_{rro}[k] = \hat{d}_{rro}[k + n_s]$ is satisfied. The parameters of (7.5) are shown in Table 7.7. The time response of $y[k]$ when $d_{rro}[k]$ is injected is shown in Figure 7.7. The line spectrum of RRO when $\hat{d}_{rro}[k] \equiv 0$ is shown in Figure 7.8.

TABLE 7.7: Parameters of RRO

i	1	2	3
Frequency $(f^i_{rro})/\mathrm{Hz}$	120	360	600
Gain (α^i_{rro})	0.15	0.03	0.006

7.3.4 Measurement Noise

The measurement noise $d_n[k]$ is defined as a white Gaussian noise with a sampling period of T_s. The standard deviation of $d_n[k]$ is $\sigma_{d_n} = 1.5 \times 10^{-2}$ of a track. The time response and spectrum of $y[k]$ when $d_n[k]$ is injected are shown in Figures 7.7 and 7.8, respectively.

7.4   Overview of the HDD Benchmark Problem Version 3

The functions in the HDD Benchmark Problem Versions 1 and 2 are unified in Version 3, and the functions and parameter files are kept separately to enhance usability. The parameter files which define the plant and disturbances of the HDD Benchmark Problem Version 2 are loaded by default. The parameter files of the HDD Benchmark Problem Version 1 are also included in the archive of the HDD Benchmark Problem Version 3.

TABLE 7.8: m-files

MATLAB function	Description
SetPlantModel.m	Define plant parameters
SetDistParam.m	Define disturbance parameters
SetForceDist.m	Generate force disturbance
SetFlutterDist.m	Generate flutter disturbance
SetRRO.m	Generate RRO
SetSensorNoise.m	Generate sensor noise

The main important functions and script files in the HDD Benchmark Problem Version 3 are shown in Table 7.8. Using these functions, the plant and disturbances can be defined by

```
% Read parameter files and define hard disk model
hdd_plantparam_file = 'hdd_plantparam';
hdd_distparam_file  = 'hdd_distparam';

% Set plant models
PlantData    = SetPlantModel(hdd_plantparam_file);

% Set disturbance parameters
DistParam    = SetDistParam(PlantData,hdd_distparam_file);

% Disturbances
ForceDist    = SetForceDist(PlantData,DistParam);
FlutterDist  = SetFlutterDist(PlantData,DistParam);
SensorNoise  = SetSensorNoise(PlantData,DistParam);
RRO          = SetRRO(PlantData,DistParam);
```

In this script, `hdd_plantparam(.m)` and `hdd_distparam(.m)` are m-files that define the parameters of the plant and the disturbances, and `SetPlantModel` and `SetDistParam` execute these files by searching for the names in the MATLAB® path. If the parameters of the plant and/or the disturbances need to be changed, these parameters files will have to be located in the current working directory before applying any changes. The variables PlantData, DistParam, ForceDist, FlutterDist, SensorNoise, and RRO generated by this script are struct variables. The specifications of the main functions are shown below.

SetPlantModel.m

Function	Define the plant model	
Syntax	[PlantData] = SetPlantModel(fname,options)	
Input arg.	fname	file name of plant parameter file (m-file)
	options.seekmode	0:following, 1:seeking
Output arg.	PlantData.Pn	Reduced-order model P_n
	PlantData.Pf	Full-order model P_f
	PlantData.Pfpert	Full-order perturbed model
	PlantData.m	Equivalent mass m
	PlantData.Tp	Track pitch T_p
	PlantData.Kf	Force constant K_f
	PlantData.Kg	Input gain $K_g = K_f/m$
	PlantData.Kp	Gain $K_p = K_f/(mT_p)$
	PlantData.rpm	Rotation speed [rpm]
	PlantData.num_servo	Sector number
	PlantData.Ts	Sampling period T_s of PES (Measurement output)
	PlantData.Ty	Step size of $y(t)$ (for simulation)
	PlantData.Tu	Step site of $u(t)$ (for simulation)
	PlantData.Td	Input delay T_d
	PlantData.MechFn	Resonance frequency f_i
	PlantData.MechZeta	Damping ratio ζ_i
	PlantData.MechGain	Residue A_i
	PlantData.PnMech	Reduced-order model of P_{mech}
	PlantData.PfMech	Full-order model of P_{mech}
	PlantData.PfpertMech	Full-order perturbed model of P_{mech}
	PlantData.DeltaKp	Range of loop gain change

SetDistParam.m

Function	Define disturbance and noise parameters	
Syntax	`[DistParam] = SetDistParam(PlantData,fname)`	
Input arg.	`PlantData`	Output of `SetPlantModel`
	`fname`	File name of disturbance parameter file (m-file)
Output arg.	`DistParam.num_sim_revolution`	Length of disturbance (reduced to disk rotation)
	`DistParam.AmpSensorNoise`	STD σ_{dn} of measurement noise
	`DistParam.AmpForceDist`	STD σ_{dt} of torque disturbance
	`DistParam.AmpFlutterDist`	Gain α^i_{flt} of $G_{flt}(s)$
	`DistParam.FlutterFreq`	Resonance frequency f^i_{flt} of $G_{flt}(s)$
	`DistParam.FlutterFreqZeta`	Damping ratio ζ^i_{flt} of $G_{flt}(s)$
	`DistParam.FreqRRO`	Frequency f^i_{rro} of RRO
	`DistParam.AmpRRO`	Amplitude α^i_{rro} of RRO
	`DistParam.RROSequence`	Time series data $\hat{d}_{rro}[k]$ of RRO
	`DistParam.Seed_ForceDist`	Seed of random number for torque disturbance
	`DistParam.Seed_SensorNoise`	Seed of random number for measurement noise
	`DistParam.Seed_FlutterDist`	Seed of random number for flutter disturbance
	`DistParam.Seed_RRODist`	Seed of random number for phase of RRO

SetForceDist.m

Function	Generate force disturbance	
Syntax	`[ForceDist] = SetForceDist(PlantData,DistParam)`	
Input arg.	`PlantData`	Output of `SetPlantModel`
	`DistParam`	Output of `SetDistParam`ĆÌŘoÛÍṬÏŘŤ
Output arg.	`ForceDist.Data`	Time series data of force disturbance
	`ForceDist.Time`	Time vector of `ForceDist.Data`
	`ForceDist.Ts`	Sampling period of `ForceDist.Data`
	`ForceDist.DataAtPes`	Time series data of force disturbance at PES
	`ForceDist.TimeAtPes`	Time vector of `ForceDist.DataAtPes`
	`ForceDist.TsAtPes`	Sampling period of `ForceDist.DataAtPes`
	`ForceDist.Spec`	Spectrum of force disturbance
	`ForceDist.SpecAtPes`	Spectrum of force disturbance at PES
	`ForceDist.Freq`	Frequency vector

272

`SetFlutterDist.m`

Function	Generate flutter disturbance	
Syntax	`[FlutterDist] = SetFlutterDist(PlantData,DistParam)`	
Input arg.	`PlantData`	Output of `SetPlantModel`
	`DistParam`	Output of `SetDistParam`
Output arg.	`FlutterDist.Data`	Time series data of flutter disturbance
	`FlutterDist.Time`	Time vector of `FlutterDist.Data`
	`FlutterDist.Ts`	Sampling period of `FlutterDist.Data`
	`FlutterDist.DataAtPes`	Time series data of flutter disturbance at
	`FlutterDist.TimeAtPes`	Time vector of `FlutterDist.DataAtPes`
	`FlutterDist.TsAtPes`	Sampling period of `FlutterDist.DataAtPes`
	`FlutterDist.Spec`	Spectrum of flutter disturbance at PES
	`FlutterDist.Freq`	Frequency vector

`SetRRO.m`

Function	Generate RRO	
Syntax	`[RRO] = SetRRO(PlantData,DistParam)`	
Input arg.	`PlantData`	Output of `SetPlantModel`
	`DistParam`	Output of `SetDistParam`
Output arg.	`RRO.Data`	Time series data of RRO
	`RRO.Time`	Time vector of `RRO.Data`
	`RRO.Ts`	Sampling period of `RRO.Data`
	`RRO.Spec`	Spectrum of RRO
	`RRO.Freq`	Frequency vector

`SetSensorNoise.m`

Function	Generate sensor noise	
Syntax	`[SensorNoise] = SetSensorNoise(PlantData,DistParam)`	
Input arg.	`PlantData`	Output of `SetPlantModel`
	`DistParam`	Output of `SetDistParam`
Output arg.	`SensorNoise.Data`	Time series data of sensor noise
	`SensorNoise.Time`	Time vector of `SensorNoise.Data`
	`SensorNoise.Ts`	Sampling period of `SensorNoise.Data`
	`SensorNoise.Spec`	Spectrum of sensor noise
	`SensorNoise.Freq`	Frequency vector

7.5 Example of Controller Design

In the technical meetings on Industrial and Control, IEEJ, special sessions for track-following control and track-seeking control were held in September 2005 and October 2006, respectively. In this section, the problem formulation and solutions presented at these special sessions are introduced. Design examples included in the HDD Benchmark Problem Version 3 are also introduced and discussed.

7.5.1 Track-Following Control Problem

The main objective of track-following control design is to design a feedback controller such that the 3σ value of Position Error Signal (PES) is as small as possible for both the nominal and perturbed plants in the presence of disturbances. Note that the sampling period of PES is fixed at T_s, but the update rate of the control input can be chosen to be arbitrarily small. Under these assumptions, the points to be evaluated are

1. stability of the closed-loop system for all the nominal and perturbed models, including the change in loop gain;

2. 3σ value of Non-Repeatable Run-Out (NRRO) of PES for all the nominal and perturbed models, including the change in loop gain; and

3. peak-to-peak value of RRO of PES for all the perturbed models, including the change in loop gain.

It is worth noting that RRO and NRRO are evaluated separately as their characteristics are different.

Next, the control performance obtained by a Proportional-Integral-Derivative (PID) control with a multi-rate notch filter (included in the HDD Benchmark Problem Version 3) is introduced. In this example, the plant parameter file in Version 1 is used and the block diagram of the control system is shown in Figure 7.9. The sampling period of the PID controller K_{pid} is T_s and that of the notch filter K_{nc} is $\frac{T_s}{2}$. The notch filter K_{nc} is composed of four notch filters to remove frequency components around 4.1 kHz, 5 kHz, 7 kHz, and 12.3 kHz. The Bode plots of K_{pid} and K_{nc} are shown in Figure 7.10. The up-sampler in Figure 7.9 interpolates the output using a ZOH.

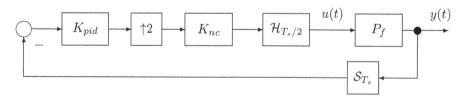

FIGURE 7.9: Block diagram of control system for track-following.

The time response of PES is shown in Figure 7.11. This figure is obtained by the program provided in the HDD Benchmark Problem, with the vertical and horizontal axes showing PES and sector number, respectively. It should be noted that the PES is shown in percentage of a track. In this figure, the

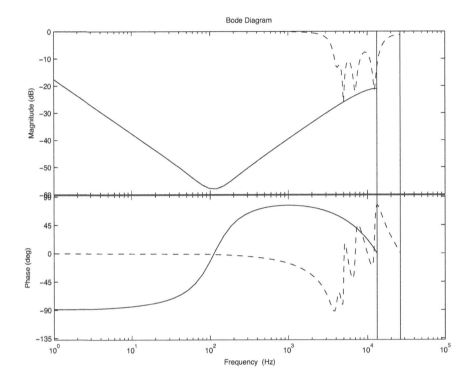

FIGURE 7.10: Bode plots of PID controller and notch filter.

6σ value of NRRO of PES, maximum and minimum values of PES, maximum and minimum values of NRRO of PES, and RRO of PES are displayed.

Figures 7.12(a) and 7.12(b) show the 3σ values of NRRO and RRO of PES obtained by track-following simulations for thirty perturbed models. Ten of the perturbed models are generated by changing the resonant frequencies, damping ratios, and residues of mechanical vibration modes. Loop gain changes of $\pm10\%$ are then considered for the ten perturbed models to generate another twenty perturbed models.

7.5.2 Track-Seeking Control Problem

In the track-seeking control problem, the following assumptions are made:

1. no disturbances are taken into account;

2. the magnitude of the control input is not limited;

3. two seek distances of one track and ten tracks are considered; and

4. the seeking time is defined as the time in which the head position settles

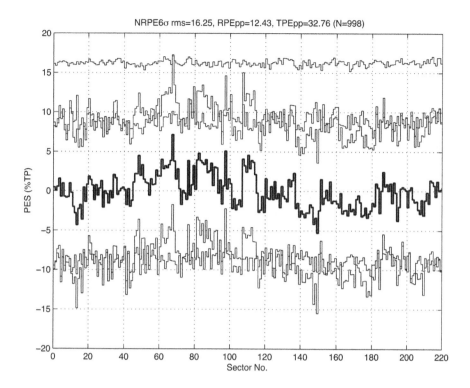

FIGURE 7.11: PES vs sector number.

within ± 0.1 of track width of the target track for all the nominal and perturbed models.

The main objective of the track-seeking control problem is to design a track-seeking control system which achieves the shortest seeking time for both one and ten track-seeks. The controller parameters can be optimized for both the one track and ten track seeks as long as the same control method is used. In addition to the seek time, the following points are also evaluated:

1. Overshoot of the head. An overshoot larger than 0.5 of a track is considered undesirable;

2. Magnitude and the smoothness of the control input. The control input has to be reasonably smooth;

3. Spectrum of the residual vibrations. The residual vibrations are evaluated from the end of the seek to $t = 10$ ms.

Next, the simulation results of the Final-State Control (FSC) included in the HDD Benchmark Problem Version 3 are introduced. In this example, three kinds of feedforward inputs, namely, `ffdata0.mat`, `ffdata1.mat`, and `ffdata2.mat`, are provided. `ffdata0.mat` is a minimum jerk input, and `ffdata1.mat` and `ffdata2.mat` are the Frequency-shaped Final-State Control (FFSC) control inputs in which the spectrum around resonant modes is reduced. These feedforward inputs were designed based on the plant parameters used in Version 2.

Last, the results of the track-seeking simulations using `ffdata0.mat` and `ffdata2.mat` are presented. Figure 7.13 shows the time responses of the feedforward inputs where the dashed and solid lines show `ffinput0.mat` and `ffinput2.mat`, respectively. The time response of `ffinput2.mat` is smoother than that of `ffinput0.mat`. However, the maximum absolute value of `ffinput2.mat` is larger than that of `ffinput0.mat`.

The track-seeking responses when `ffinput0.mat` and `ffdata2.mat` are injected into the perturbed models are shown in Figures 7.14(a) and 7.14(b), respectively. It should be noted that the results when the loop gain is changed by ±10% are not shown here as the feedback controller was not designed in this simulation. From the simulation results, it can be verified that the FFSC can achieve a faster track-seeking response as compared to that using the minimum jerk input.

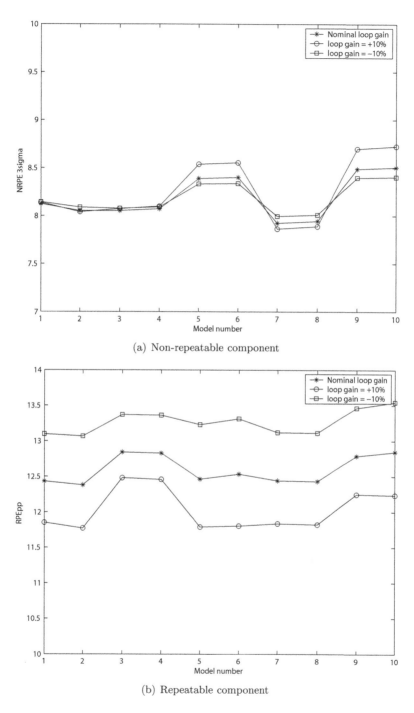

(a) Non-repeatable component

(b) Repeatable component

FIGURE 7.12: Tracking errors.

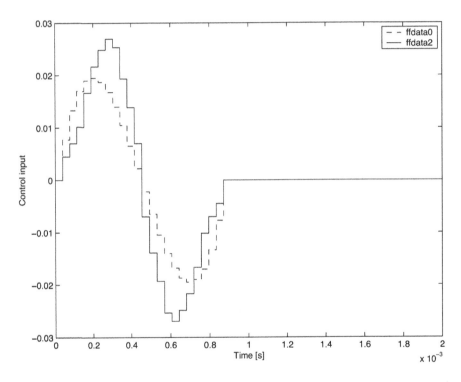

FIGURE 7.13: Time responses of feedforward inputs.

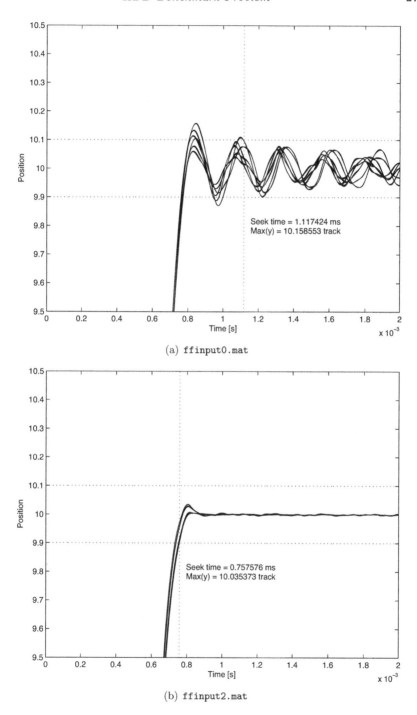

(a) `ffinput0.mat`

(b) `ffinput2.mat`

FIGURE 7.14: Track-seeking responses.

Bibliography

[1] T. Hara, "Designing Multi-Rate Controllers for the HDD Benchmark Problem," *The Papers of Technical Meeting on Industrial Instrumentation and Control, IEEJ*, IIC-05-109, pp. 27–32, 2005-9 (in Japanese).

[2] T. Atsumi, "Designing a Following Controller by Using Vector Locus in Hard Disk Drives," *The Papers of Technical Meeting on Industrial Instrumentation and Control, IEEJ*, IIC-05-110, pp. 33–38, 2005-9 (in Japanese).

[3] M. Hirata and Y. Hasegawa, "H_∞ Controller Design for Hard Disk Benchmark Problem," *The Papers of Technical Meeting on Industrial Instrumentation and Control, IEEJ*, IIC-05-111, pp. 39–42, 2005-9 (in Japanese).

[4] H. Fujioka, "A Design for HDD Benchmark Problem with Sampled-Data Control Toolbox," *The Papers of Technical Meeting on Industrial Instrumentation and Control, IEEJ*, IIC-05-112, pp. 43–46, 2005-9 (in Japanese).

[5] N. Ishida, M. Kawafuku, M. Iwasaki, H. Hirai, M. Kobayashi, and A. Okuyama, "Rejection Control for Repeatable Runout of HDDs Using Inverse Characteristic Based on Frequency Analysis— Investigation into Benchmark Model," *The Papers of Technical Meeting on Industrial Instrumentation and Control, IEEJ*, IIC-05-112, pp. 47–60, 2005-9 (in Japanese).

[6] K. Fukushima and H. Fujimoto, "Short Span Seeking with Vibration Suppression PTC for Benchmark Problem," *The Papers of Technical Meeting on Industrial Instrumentation and Control, IEEJ*, IIC-06-139, pp. 43–46, 2006-9, (in Japanese).

[7] M. Hirata and Y. Hasegawa, "FFSC Design for Hard Disk Benchmark Problem," *The Papers of Technical Meeting on Industrial Instrumentation and Control, IEEJ*, IIC-06-140, pp. 47–50, 2006-9 (in Japanese).

[8] M. Yamamoto, K. Ito, M. Kawafuku, and M. Iwasaki, "Application of Minimum Jerk Control to Benchmark Problem Short Span Seeking," *The Papers of Technical Meeting on Industrial Instrumentation and Control, IEEJ*, IIC-06-141, pp. 51–56, 2006-9 (in Japanese).

[9] N. Nakamura, Y. Hori, and N. Bando, "Performance Verification of Self Servo Track Writer Using Hard Disk Benchmark Software," *The Papers of Technical Meeting on Industrial Instrumentation and Control, IEEJ*, IIC-06-142, pp. 57–60, 2006-9 (in Japanese).

[10] T. Sato and S. Masuda, "Head-Positioning Control in a Hard Disk Drive by Using Generalized Predictive Control Considering Intersample," *The Papers of Technical Meeting on Industrial Instrumentation and Control, IEEJ*, IIC-06-143, pp. 61–64, 2006-9 (in Japanese).

Index